Girlwonder

Girlwonder

HOLLY HARTMAN

with the Editors of

INFORMATION PLEASE®

Information
Please®
www.infoplease.com

Houghton Mifflin Company
Boston • New York
2003

For Cynthia

First Houghton Mifflin paperback edition 2003
Copyright © 2003 by Pearson Education, Inc.
Information Please® and Information Please Almanac®
are registered trademarks of Pearson Education, Inc.
Visit our Web site at www.infoplease.com.

For information about permission to reproduce selections from
this book, write to Permissions, Houghton Mifflin Company,
215 Park Avenue South, New York, New York 10003.

Visit our Web site: www.houghtonmifflinbooks.com.

Library of Congress Cataloging-in-Publication Data
Girlwonder : every girl's guide to the fantastic feats, cool qualities, and
remarkable abilities of women and girls / Holly Hartman (editor).
p. cm.
An updated and expanded edition of: The information please
girls' almanac / Margo McLoone and Alice Siegel, 1995.
ISBN 0-618-31939-5
1. Teenage girls—Life skills guides—Juvenile literature. 2. Socialization—
Juvenile literature. 3. Teenage girls—Psychology—Juvenile literature.
4. Adolescent psychology—Juvenile literature. 5. Women—United
States—Biography—Juvenile literature. I. Hartman, Holly. II. McLoone,
Margo. Information please girls' almanac.
HQ798.G58 2003
305.235—dc21   2003047897

Printed in the United States of America

Text design by Joyce C. Weston

QUM 10 9 8 7 6 5 4 3 2 1

This book contains information on a broad range of subjects. None
of the information is intended as a substitute for the advice of a professional in each of the relevant areas. The reader should consult a
professional in connection with matters that may require professional
or expert attention.

CONTENTS

Calendar

JANUARY

Gemstone: garnet
Flower: snowdrop

1 On this day in 1892, Annie Moore, a fifteen-year-old from Ireland, became the first of twelve million immigrants to pass through New York City's Ellis Island Immigration Station.

2 On this day in 1991, Sharon Pratt Dixon was sworn in as mayor of Washington, D.C., becoming the first black woman to serve as mayor of a major U.S. city.

3 On this day in 1987, American soul singer Aretha Franklin became the first woman inducted into the Rock and Roll Hall of Fame.

4 On this day in 1991, twelve-year-old Fu Mingxia of China became the youngest winner ever at the World Swimming Championships.

5 On this day in 1925, Nellie Tayloe Ross of Wyoming became the first female state governor.

6 Joan of Arc, French soldier and saint, was born on this day in 1412.

7 On this day in 1955, American singer Marian Anderson made her debut with New York's Metropolitan Opera.

8 Emily Greene Balch, Nobel Prize–winning American peace activist, was born on this day in 1867.

9 Carrie Chapman Catt, American founder of the League of Women Voters, was born on this day in 1859.

10 On this day in 1917, the National Woman's Party picketed the White House with banners such as "Mr. President, What Will You Do for Woman Suffrage?"

11 On this day in 1935, American aviator Amelia Earhart began her landmark flight across the Pacific Ocean. She was the first aviator to make that journey solo.

12 On this day in 1932, Hattie Caraway of Arkansas became the first woman elected to the U.S. Senate.

13 On this day in 1976, American Sarah Caldwell became the first woman to conduct at New York's Metropolitan Opera House.

14 On this day in 1970, Americans Diana Ross and the Supremes performed their last concert together, in Las Vegas.

15 On this day in 1559, England's Queen Elizabeth I was crowned.

16 On this day in 1978, NASA selected its first female astronauts: Anna Fisher, Shannon Lucid, Judith Resnick, Sally Ride, Margaret Seddon, and Kathryn Sullivan.

17 On this day in 1893, Queen Liliuokalani of Hawaii was forced from her throne by a group of American businessmen.

18 On this day in 1986, Ann Bancroft of Minnesota became the first woman to walk to the North Pole.

19 On this day in 1966, Indira Gandhi was elected prime minister of India.

20 On this day in 1952, Patricia McCormick, the first female professional bullfighter from the United States, made her debut in Juárez, Mexico.

21 American actress Geena Davis was born on this day in 1957.

22 American soprano Rosa Ponselle was born on this day in 1897.

23 On this day in 1849, Elizabeth Blackwell, the first female doctor in the United States, graduated first in her class from New York's Geneva College.

24 On this day in 1985, Penny Harrington became the first female police chief of a major U.S. city, in Portland, Oregon.

25 On this day in 1890, American journalist Elizabeth Cochrane Seaman, who used the pen name Nellie Bly, completed a trip around the world in a record 72 days, 6 hours, and 11 minutes.

26 American architect Julia Morgan was born on this day in 1872.

27 On this day in 1961, American soprano Leontyne Price made her debut with New York's Metropolitan Opera.

28 On this day in 1908, Julia Ward Howe, author of "The Battle Hymn of the Republic," became the first woman elected to the American Academy of Arts and Letters.

29 Oprah Winfrey, American author and talk show host, was born on this day in 1954.

30 British actress Vanessa Redgrave was born on this day in 1937.

31 Russian prima ballerina Anna Pavlova was born on this day in 1881.

FEBRUARY

Gemstone: amethyst
Flower: violet

1 On this day in 1978, a postage stamp honoring abolitionist Harriet Tubman was issued. She was the first African American woman to have her likeness on a U.S. postage stamp.

2 American writer Ayn Rand was born on this day in 1905.

3 On this day in 1977, Chicago secretary Iris Rivera lost her job for refusing to make coffee. Women across the city organized protests, bringing attention to problems faced by female office workers.

4 Rosa Parks, American civil rights activist, was born on this day in 1913.

5 On this day in 1931, Maxine Dunlap became the first American woman to earn a glider pilot license.

6 On this day in 2000, First Lady Hillary Rodham Clinton launched her successful campaign to become a U.S. senator from New York.

7 American author Laura Ingalls Wilder was born on this day in 1867.

8 American poet Elizabeth Bishop was born on this day in 1911.

9 Alice Walker, Pulitzer Prize–winning American author, was born on this day in 1944.

10 On this day in 1968, American Peggy Fleming won the gold medal in women's figure skating at the Winter Olympics in Grenoble, France.

11 On this day in 1989, in Boston, the Reverend Barbara C. Harris became the first woman consecrated as a bishop in the Episcopal Church.

12 American author Judy Blume was born on this day in 1938.

13 Elaine Pagels, American religious scholar, was born on this day in 1943.

14 On this day in 1920, the League of Women Voters was founded in Chicago.

15 On this day in 1946, American Edith Houghton became the first woman hired as a Major League Baseball scout.

16 On this day in 1926, tennis legends Helen Wills of the United States and Suzanne Lenglen of France faced off for the first and only time. Lenglen won.

17 Marian Anderson, American contralto and opera legend, was born on this day in 1897.

18 Toni Morrison, Nobel Prize–winning American novelist, was born on this day in 1931.

19 American author Carson McCullers was born on this day in 1917.

20 On this day in 1998, fifteen-year-old American Tara Lipinski became the youngest Olympic gold medal winner in women's figure skating, at the Winter Olympics in Nagano, Japan.

21 On this day in 1866, American Lucy Hobbs became the first woman to graduate from dental school, at the Ohio College of Dental Surgery.

22 On this day in 1992, American Kristi Yamaguchi won the gold medal in women's figure skating at the Winter Olympics in Albertville, Canada.

23 American aviator Ruth Nichols was born on this day in 1901.

24 On this day in 1999, American Lauryn Hill set a record by winning five Grammy Awards for her album *The Miseducation of Lauryn Hill.*

25 Adele Davis, American nutrition expert, was born on this day in 1905.

26 British actress Margaret Leighton was born on this day in 1922.

27 On this day in 1998, Britain's House of Lords agreed that the country's throne would pass to a monarch's first-born child, regardless of gender.

28 On this day in 1998, Celine Dion's "My Heart Will Go On," from the movie *Titanic,* topped the charts.

29 Mother Ann Lee, founder of the Shaker religion, was born in England on this day in 1736.

MARCH

Gemstone: aquamarine
Flower: daffodil

1 Today begins Women's History Month, a monthlong national celebration in the United States since 1987.

2 On this day in 1878, one of the first women's hotels in the United States, the Barbizon, opened in New York City.

3 On this day in 1887, American Anne Sullivan became the teacher of Helen Keller, a six-year-old girl in Alabama who was deaf and blind.

4 Jane Goodall, British ethologist and famed observer of chimpanzees, was born on this day in 1934.

5 Louise Pearce, an American scientist who developed a drug to control sleeping sickness, was born on this day in 1885.

6 British poet Elizabeth Barrett Browning was born on this day in 1806.

7 On this day in 1870, in Wyoming, women were allowed to serve on a grand jury for the first time in U.S. history.

8 Today is International Women's Day, a day to honor the world's working women. It was first proclaimed in 1909.

9 On this day in 1990, Dr. Antonia Novello was sworn in as U.S. surgeon general, becoming the first woman (and the first Hispanic) to hold that job.

10 Today is Harriet Tubman Day. Tubman, an American abolitionist and former slave who helped to lead hundreds of slaves to freedom, died on this day in 1913.

11 On this day in 1993, the Senate confirmed Janet Reno's appointment as U.S. attorney general. She became the first woman to hold that job.

12 Today is Girl Scout Day. The Girl Guides, the group that became the Girl Scouts of the USA, was founded on this day in 1912.

13 Abigail Fillmore, the first lady who established the White House library, was born on this day in 1798.

14 American photographer Diane Arbus was born on this day in 1923.

15 On this day in 1964, actress Elizabeth Taylor wed actor Richard Burton, beginning what would become one of the most watched celebrity marriages of all time.

16 Sarah Caldwell, the first woman to conduct at New York's Metropolitan Opera House, was born on this day in 1924.

17 On this day in 1969, Golda Meir became prime minister of Israel.

18 Queen Latifah, American rap star and actress, was born on this day in 1970.

19 Edith Nourse Rogers, who was elected eighteen times to the U.S. House of Representatives, was born on this day in 1881.

20 Today is Harriet Beecher Stowe Day. On this day in 1852, the American author's famous novel *Uncle Tom's Cabin* was published.

21 On this day in 1964, American singer Judy Collins made her Carnegie Hall debut.

22 On this day in 1893, the first women's college basketball game was played at Smith College.

23 American actress Joan Crawford was born on this day in 1908.

24 On this day in 1603, Queen Elizabeth I of England died, ending her forty-five-year reign, which is known as the Golden Age or the Elizabethan era.

25 Aretha Franklin, American singer known as "the Queen of Soul," was born on this day in 1942.

26 Sandra Day O'Connor, first female justice on the U.S. Supreme Court, was born on this day in 1930.

27 Sarah Vaughan, American jazz singer, was born on this day in 1924.

28 American actress Dianne Wiest was born on this day in 1948.

29 On this day in 1994, eleven-year-old Anna Paquin of New Zealand won the Oscar for best supporting actress for her role in the movie *The Piano*. She had never acted before.

30 French-Canadian singer Celine Dion was born on this day in 1968.

31 On this day in 1776, Abigail Adams wrote a letter to her husband, John, a member of the Continental Congress, and asked him to "remember the Ladies and be more generous and favorable than your ancestors were."

APRIL

Gemstone: diamond
Flower: sweet pea

1 Florence Blanchfield, a nurse who was the first woman to become an officer in the U.S. Army, was born on this day in 1884.

2 American country singer Emmylou Harris was born on this day in 1947.

3 Doris Day, American actress and singer, was born on this day in 1924.

4 On this day in 1887, Susanna Medora Salter became the first woman elected mayor of an American town, in Argonia, Kansas.

5 Sybil Ludington was born on this day in 1761. She was a Revolutionary War heroine who, at age sixteen, rode her horse forty miles at night to summon reinforcements for the colonists.

6 American actress Marilu Henner was born on this day in 1952.

7 On this day in 1987, the National Museum of Women in the Arts opened in Washington, D.C.

8 Mary Pickford, Academy Award–winning American actress and producer, was born on this day in 1893.

9 On this day in 1939, singer Marian Anderson gave a history-making performance at the Lincoln Memorial. Originally scheduled to perform at Constitution Hall, Anderson was booted from that venue because she was African American.

10 On this day in 2001, Jane Swift became the first female governor of Massachusetts.

11 On this day in 1953, Oveta Culp Hobby became the first woman to serve as secretary of health, education, and welfare. She was also the first director of the Women's Auxiliary Army Corps (WAAC) and the first woman to receive the U.S. Army's Distinguished Service Medal.

12 Beverly Cleary, American author best known for her series of books about Ramona Quimby and Henry Huggins, was born on this day in 1916.

13 American Anne Sullivan, famed teacher of Helen Keller, was born on this day in 1866.

14 Loretta Lynn, American country music singer and songwriter, was born on this day in 1935.

15 Bessie Smith, American blues and jazz singer known as "the Empress of the Blues," was born on this day in 1894.

16 On this day in 1912, American aviator Harriet Quimby became the first woman to fly across the English Channel.

17 Sirimavo Bandaranaike, the first woman to serve as prime minister of Sri Lanka, was born on this day in 1960.

18 On this day in 1956, American actress Grace Kelly married Prince Rainier of Monaco.

19 American movie star Jayne Mansfield was born on this day in 1933.

20 American actress Jessica Lange was born on this day in 1949.

21 Charlotte Brontë, British author of the novel *Jane Eyre*, was born on this day in 1816.

22 Queen Isabella I of Spain was born on this day in 1451.

23 Shirley Temple Black, American child star who became a diplomat, was born on this day in 1928.

24 American actress Shirley Maclaine was born on this day in 1934.

25 Ella Fitzgerald, American blues and jazz singer known for her style of "scat" singing, was born on this day in 1917.

26 Anita Loos, American screenwriter and author of the novel *Gentlemen Prefer Blondes*, was born on this day in 1893.

27 American civil rights activist Coretta Scott King was born on this day in 1927.

28 American writer Harper Lee, whose only novel, *To Kill a Mockingbird,* won the Pulitzer Prize, was born on this day in 1926.

29 On this day in 1925, Florence Sabin, who researched the origin of blood cells, became the first woman elected to the National Academy of Sciences. She was also the first woman to graduate from Johns Hopkins University School of Medicine.

30 American actress Cloris Leachman was born on this day in 1926.

MAY

Gemstone: emerald
Flower: lily of the valley

1 Mary Harris Jones, known as "Mother Jones," was born on this day in 1830. She organized American coal miners to strike for better wages and working conditions.

2 On this day in 1970, Diane Crump became the first female jockey to ride in the Kentucky Derby.

3 On this day in 1899, the Women Lawyers' Association was formed to promote the interests of women in the law nationwide.

4 Audrey Hepburn, American actress and children's rights advocate, was born on this day in 1929.

5 Nellie Bly, American journalist who gained fame for the record-breaking speed of her journey around the world, was born on this day, probably in 1867.

6 On this day in 1978, the first jazz festival for female musicians was held in Kansas City, Kansas. Marian McPartland and Betty Carter were among the performers.

7 On this day in 1789, Martha Washington hosted the first presidential inaugural ball in New York City.

8 On this day in 1999, Nancy Ruth Mace became the first female cadet to graduate from the Citadel, the formerly all-male military school in South Carolina.

9 Belle Boyd, American actress and Civil War spy, was born on this day in 1844.

10 On this day in 1908, the first Mother's Day celebrations in the United States took place in Philadelphia and in Grafton, West Virginia.

11 American dancer and choreographer Martha Graham was born on this day in 1894.

12 Florence Nightingale, British nurse known for her work in the Crimean War, was born on this day in 1820.

13 Maria Theresa, empress of Austria, was born on this day in 1717. A wise and able ruler, she unified the Austro-Hungarian Empire.

14 On this day in 1942, the Women's Auxiliary Army Corps (WAAC) was established in the United States.

15 On this day in 1851, the first sorority was founded by sixteen women at Wesleyan College in Macon, Georgia. First named the Adelphians, they later became Alpha Delta Pi.

16 Pop singer Janet Jackson was born on this day in 1966.

17 On this day in 1978, women made their debut as White House honor guards during a welcoming ceremony for President Kenneth Kaunda of Zambia.

18 On this day in 1953, Jacqueline Cochran became the first woman to break the sound barrier when she flew an F-86 over Roger's Dry Lake, California, at the speed of 652.337 miles per hour. Eleven years later, she flew at a speed of 1,429.2 miles per hour, more than twice the speed of sound.

19 On this day in 1930, playwright Lorraine Hansberry was born. Her play *A Raisin in the Sun* (1959) was the first play by a black woman to be produced on Broadway.

20 On this day in 1932, American aviator Amelia Earhart began the first solo flight by a woman across the Atlantic Ocean. She took a can of tomato juice and a thermos of soup for her 14-hour, 56-minute flight from Newfoundland to Ireland.

21 On this day in 1881, nurse Clara Barton founded the American Red Cross. She worked for ten years to establish this foundation, which provides peacetime disaster relief as well as wartime medical assistance.

22 American impressionist painter Mary Cassatt was born on this day in 1884.

23 Margaret Wise Brown, American author of such children's books as *Goodnight Moon,* was born on this day in 1910.

24 Queen Victoria, who ruled Great Britain for more than sixty years, was born on this day in 1819.

25 American opera singer Beverly Sills was born on this day in 1929.

26 American astronaut Sally Ride was born on this day in 1951. She became the first American woman to travel in space.

27 American poet Julia Ward Howe, who wrote "The Battle Hymn of the Republic," was born on this day in 1819.

28 The Dionne quintuplets were born in Canada on this day in 1934. These five girls — Annette, Cecile, Emilie, Marie, and Yvonne — were the first quints to live to adulthood.

29 On this day in 1921, American novelist Edith Wharton became the first woman to win a Pulitzer Prize for fiction. She won the award for her novel *The Age of Innocence.*

30 Today Joan of Arc Feast Day is celebrated in France. Joan of Arc led the French army against the invading English army in 1429. She was captured and burned at the stake on this day in 1431.

31 Patricia Harris, the first African American woman to serve in a presidential cabinet, was born on this day in 1924. She served as secretary of both the Department of Housing and Urban Development and the Department of Health and Human Services under President Jimmy Carter.

JUNE

Gemstone: pearl
Flower: rose

1 American actress Marilyn Monroe was born Norma Jean Baker on this day in 1926.

2 On this day in 1953, Queen Elizabeth II of Great Britain was crowned in London.

3 On this day in 1972, Sally Jean Priesand was ordained as the first female rabbi in the United States.

4 Catharine McCulloch, who in 1898 became the first woman to argue a case before the U.S. Supreme Court, was born on this day in 1862.

5 American anthropologist Ruth Benedict was born on this day in 1887.

6 On this day in 1872, American suffragist Susan B. Anthony was arrested and fined for voting in an election — at that time, it was illegal for a woman to do so.

7 Poet Gwendolyn Brooks, the first African American woman to win a Pulitzer Prize for poetry, was born on this day in 1917.

8 American comedienne Joan Rivers was born on this day in 1933.

9 On this day in 1993, Masako Owada married Crown Prince Naruhito of Japan.

10 On this day in 1692, Bridget Bishop of Salem, Massachusetts, became the first of the Salem "witches" to be hanged.

11 Jeanette Rankin, the first woman elected to the U.S. Congress, was born on this day in 1880. She fought for women's right to vote and was an ardent pacifist who, at age eighty-seven, led a protest against the Vietnam War known as the Jeanette Rankin Brigade.

12 Anne Frank, who is remembered for the diary she wrote while hiding from the Nazis in Amsterdam during World War II, was born on this day in 1929.

13 On this day in 1893, the first British Ladies Golf Championship was held in England.

14 Tennis champion Steffi Graf was born in West Germany on this day in 1969. She became a professional player at age thirteen.

15 American sculptor Malvina Hoffman was born on this day in 1887.

16 Mary Katherine Goddard, publisher of three colonial newspapers and the first woman to serve as a postmaster in the United States, was born on this day in 1738.

17 On this day in 1885, the Statue of Liberty, the most famous American statue of a woman, arrived in New York City aboard a French ship.

18 On this day in 1983, Sally Ride became the first American woman in space when she began a six-day mission aboard the space shuttle *Challenger*.

19 On this day in 1984, Geraldine Ferraro, a Democratic congresswoman from Queens, New York, became the first woman nominated by a major political party as candidate for vice president.

20 American author and playwright Lillian Hellman was born on this day in 1905.

21 On this day in 1913, American daredevil Georgia Broadwick became the first woman to jump out of an airplane. (She wore a parachute, of course.)

22 American author and aviator Anne Morrow Lindbergh was born on this day in 1906.

23 Championship African American runner Wilma Rudolph, who in 1960 became the first woman to win three gold medals at a single Olympics, was born on this day in 1940.

24 On this day in 1647, Margaret Brent appeared before the Maryland assembly demanding that women be granted the right to vote. She was the first woman in Maryland to own property and one of the first known suffragists in American history.

25 Singer and songwriter Carly Simon was born on this day in 1945.

26 Mildred "Babe" Didrikson Zaharias, American all-around star athlete, was born on this day in 1913.

27 Emma Goldman, Russian-born American anarchist, was born on this day in 1869. She became a lecturer on free speech and women's rights and was considered the most uncompromising radical of her time.

28 On this day in 1904, Helen Keller, who had been blind and deaf since infancy, graduated from Radcliffe College with honors.

29 Politician Elizabeth Dole was born on this day in 1936. She served as a cabinet member under both Ronald Reagan and George H. W. Bush, subsequently served as president of the American Red Cross, and was elected a U.S. senator from North Carolina in 2002.

30 On this day in 1936, American author Margaret Mitchell published her best-selling novel *Gone With the Wind*.

JULY

Gemstone: ruby
Flower: water lily

1 Diana Spencer, whose marriage to Prince Charles of Great Britain would make her the Princess of Wales, was born on this day near Sandringham, England, in 1961.

2 On this day in 1979, a U.S. dollar coin that pictured women's rights advocate Susan B. Anthony was released.

3 M. F. K. (Mary Frances Kennedy) Fisher, American food writer, was born on this day in 1908.

4 On this day in 1960, Liza Redfield became the first woman to conduct an orchestra for a Broadway play. She led a twenty-four-piece orchestra in *The Music Man*.

5 On this day in 1972, Juanita Kreps became the first female director of the New York Stock Exchange. She later became the first woman appointed secretary of commerce.

6 On this day in 1967, Althea Gibson became the first African American tennis player to win a singles title at Wimbledon.

7 On this day in 1946, Italian-born Mother Frances Xavier Cabrini was canonized as the first American saint.

8 On this day in 1911, Nan Jane Aspinwall arrived in New York City, completing a 4,500-mile solo horseback trip across the United States.

9 British romance novelist Barbara Cartland, the author of more than seven hundred books, was born on this day 1901.

10 Mary McLeod Bethune, the educator who founded Bethune-Cookman College in Florida and served as a presidential adviser, was born on this day in 1875. She grew up on a South Carolina plantation, where she picked cotton all day and attended school between picking seasons.

11 American actress Sela Ward was born on this day in 1956.

12 On this day in 1912, the French movie *Queen Elizabeth,* the first foreign film shown in the United States, had its stateside premiere. Sarah Bernhardt played the title role.

13 On this day in 1954, American singer Kitty Kallen's rendition of "Little Things Mean a Lot" topped the charts.

14 Emmeline Pankhurst, a leader in the British women's suffrage movement, was born on this day in 1848.

15 American singer Linda Ronstadt was born on this day in 1946.

16 Mary Baker Eddy, founder of the Christian Science religion, was born on this day in 1821.

17 Berenice Abbott, an American photographer who is best remembered for her black-and-white photographs of New York City, was born on this day in 1898.

18 On this day in 1976, Nadia Comaneci of Romania became the first gymnast ever to score a perfect 10 at the Olympic Games.

19 On this day in 1848, the first women's rights convention was held in Seneca Falls, New York.

20 Theda Bara, American silent film actress, was born on this day in 1885.

21 Louise Blanchard Bethune, the first American female architect, was born on this day in 1856.

22 Emma Lazarus, author of the poem "The New Colossus," which is inscribed on the base of the Statue of Liberty, was born on this day in 1849.

23 American supermodel Stephanie Seymour was born on this day in 1968.

24 American aviator Amelia Earhart was born on this day in 1897.

25 Louise Brown, the first test-tube baby, was born in Oldham, England, on this day in 1978. She weighed 5 pounds, 12 ounces.

26 American figure skater Dorothy Hamill, winner of the Olympic gold medal in 1976, was born on this day in 1956.

27 On this day in 1942, American singer Peggy Lee recorded her first hit record, "Why Don't You Do Right?"

28 Beatrix Potter, British author and illustrator who created the Peter Rabbit series, was born on this day in 1866.

29 Lady Diana Spencer married Prince Charles, heir to the throne of England, on this day in 1981. The ceremony was shown on television around the world.

30 Emily Brontë, British author of *Wuthering Heights,* was born on this day in 1818.

31 J. K. (Joanne Kathleen) Rowling, author of the Harry Potter series, was born on this day in 1965.

AUGUST

Gemstone: onyx
Flower: poppy

1 Astronomer Maria Mitchell, the first woman elected to the American Academy of Arts and Sciences, was born on this day in 1818.

2 American film actress Myrna Loy was born on this day in 1905.

3 On this day in 1993, the U.S. Senate voted 96–3 to confirm the appointment of Ruth Bader Ginsburg to the Supreme Court. Ginsburg was the second woman to become a Supreme Court justice.

4 Record-setting American track and field athlete Mary Decker Slaney was born on this day in 1958.

5 On this day in 1924, the comic strip *Little Orphan Annie* appeared in newspapers for the first time.

6 On this day in 1926, American Gertrude Ederle became the first woman to swim across the English Channel.

7 Mata Hari, a belly dancer who became a double agent in World War I — spying for both the French and the Germans — was born in Amsterdam on this day in 1876.

8 Marjorie Kinnan Rawlings, American author of such novels as *The Yearling,* was born on this day in 1896.

9 Australian-born British author P. L. (Pamela Lyndon) Travers, best known for her Mary Poppins books, was born on this day in 1899.

10 American fashion designer Betsey Johnson was born on this day in 1942.

11 On this day in 1862, French actress Sarah Bernhardt, later known as the "Divine Sarah," made her stage debut in Paris.

12 Katherine Lee Bates, American educator and author of "America the Beautiful," was born on this day in 1859.

13 Annie Oakley, sharpshooter who performed in Buffalo Bill's Wild West Show, was born on this day in 1860.

14 On this day in 1995, Shannon Faulkner became the first female cadet at the Citadel, the formerly all-male military school in South Carolina. (However, she dropped out in her first semester.)

15 American actress Ethel Barrymore was born on this day in 1879.

16 Madonna Louise Ciccone, American pop star, was born on this day in 1958.

17 American actress Mae West was born on this day in 1892.

18 Virginia Dare, the first European baby born on American soil, was born on this day at Roanoke Island, Virginia, in 1587.

19 On this day in 1890, the Daughters of the American Revolution, a group of women who trace their lineage to ancestors who fought in the Revolution, was formed.

20 American journalist and news anchor Connie Chung was born on this day in 1946.

21 On this day in 1980, American singer Linda Ronstadt made her Broadway debut as Mabel in Gilbert and Sullivan's operetta *The Pirates of Penzance.*

22 American writer and humorist Dorothy Parker was born on this day in 1893.

23 On this day in 1902, Fannie Farmer opened her famous cooking school in Boston.

24 On this day in 1932, American aviator Amelia Earhart became the first woman to fly nonstop across the nation. She made the trip from Los Angeles to Newark, New Jersey, in about nineteen hours.

25 American tennis champion Althea Gibson was born on this day in 1927.

26 On this day in 1920, the Nineteenth Amendment to the U.S. Constitution was ratified, granting women the right to vote in national elections.

27 Mother Teresa, who devoted her life to helping the poor in India, was born Agnes Bojaxhiu on this day in 1910 in Yugoslavia.

28 Elizabeth Ann Seton, the first person born in the New World to become a saint, was born on this day in 1774.

29 Swedish movie actress Ingrid Bergman was born on this day in 1915.

30 British author Mary Wollstonecraft, best known for her novel *Frankenstein,* was born on this day in 1797.

31 Maria Montessori, the Italian educator whose philosophies launched the Montessori schools, was born on this day in 1870.

SEPTEMBER

Gemstone: sapphire
Flower: morning glory

1 On this day in 1878, Emma Nutt of Boston, Massachusetts, was hired as the first female telephone operator in the United States.

2 Lydia Kameheka Paki Liliuokalani, queen of Hawaii from 1891 to 1893, was born on this day in 1838.

3 Sarah Orne Jewett, American author best known for writing about Maine, was born on this day in 1849.

4 American actress and dancer Mitzi Gaynor was born on this day in 1930.

5 American actress Raquel Welch was born on this day in 1940.

6 Jane Addams, the first American woman to win a Nobel Peace Prize, was born on this day in 1860. She was a social reformer who lived and worked among the poor and needy at Hull House, the famous community center she founded in Chicago.

7 Anna Mary Moses, known as "Grandma Moses," was born on this day in 1860. A painter of American rural life, she began painting when she was seventy-six years old, and between the ages of one hundred and one hundred one she completed twenty-five paintings.

8 On this day in 1921, the first Miss America, sixteen-year-old Margaret Gorman from Washington, D.C., was crowned.

9 On this day in 1893, Esther Cleveland, daughter of President Grover Cleveland and Frances Cleveland, became the first baby born in the White House.

10 On this day in 1988, German tennis champion Steffi Graf won the U.S. Open and in so doing won the Grand Slam (the Australian Open, French Open, Wimbledon, and U.S. Open titles in the same season).

11 On this day in 1850, Swedish singer Jenny Lind, known as "the Swedish Nightingale," made her American singing debut.

12 On this day in 1977, Azie Taylor Morton, the first African American woman to serve as treasurer of the United States, took office.

13 On this day in 1949, the Ladies Professional Golf Association was founded in New York City.

14 On this day in 1975, Elizabeth Ann Seton was canonized, making her the first American-born saint.

15 British mystery writer Agatha Christie was born on this day in 1890.

16 American actress Lauren Bacall was born on this day in 1924.

17 Maureen Connolly, the first woman to win the Grand Slam in tennis, was born on this day in 1934.

18 Greta Garbo, a Swedish-born actress who became a movie star in the United States, was born on this day in 1905.

19 On this day in 1970, *The Mary Tyler Moore Show* made its television debut.

20 On this day in 1973, tennis player Billie Jean King won the "Battle of the Sexes" match, defeating Bobby Riggs in three straight sets at the Houston Astrodome.

21 On this day in 1897, the *New York Sun* ran an editorial to answer eight-year-old Virginia O'Hanlon's question about whether Santa Claus existed. Their famous answer? "Yes, Virginia, there is a Santa Claus."

22 Christabel Pankhurst, British feminist leader, was born on this day in 1880.

23 Victoria Woodhull, the first woman to run for the U.S. presidency, was born on this day in 1838. Her running mate in the 1872 election was the black abolitionist leader Frederick Douglass.

24 American actress Audra Lindley was born on this day in 1918.

25 Barbara Walters, American journalist and television commentator, was born on this day in 1931.

26 Olivia Newton-John, Australian actress and singer, was born on this day in 1948.

27 American actress Gwyneth Paltrow was born on this day in 1972.

28 American tennis player Alice Marble was born on this day in 1913.

29 Swedish screen siren Anita Ekberg was born on this day in 1918.

30 On this day in 1954, American actress and singer Julie Andrews made her Broadway debut in *The Boy Friend*.

OCTOBER

Gemstone: opal
Flower: calendula

1 American actress and singer Julie Andrews was born on this day in 1935.

2 American fashion designer Donna Karan was born on this day in 1948.

3 On this day in 1922, Rebecca Felton of Georgia became the first woman to serve in the U.S. Senate.

4 American actress Susan Sarandon was born on this day in 1946.

5 British actress Kate Winslet was born on this day in 1975.

6 American tennis star Helen Wills Moody was born on this day in 1905.

7 On this day in 1975, women were permitted to enroll in the U.S. military academies for the first time.

8 On this day in 1871, Mrs. O'Leary and her cow went down in history. Legend has it that the cow kicked over a barn lantern, starting the Great Chicago Fire.

9 On this day in 1930, Laura Ingalls became the first woman to cross the nation by plane, on a nine-stop journey from New York to California.

10 American actress Helen Hayes was born on this day in 1900.

11 First lady, author, diplomat, and human-rights advocate Eleanor Roosevelt was born on this day in 1884.

12 American track and field athlete Marion Jones was born on this day in 1975.

13 Mary Hays McCauley, who became famous as "Molly Pitcher," was born on this day in 1754. She fought in the Battle of Monmouth during the American Revolution.

14 Professor and political philosopher Hannah Arendt was born on this day in 1906.

15 On this day in 1860, eleven-year-old Grace Bedell sent a letter to presidential candidate Abraham Lincoln saying that he would look better if he grew a beard. So he did.

16 On this day in 1793, Queen Marie Antoinette of France was beheaded.

17 On this day in 1979, Mother Teresa was awarded the Nobel Peace Prize for her work with the poor in India.

18 Czech-born tennis great Martina Navratilova was born on this day in 1956.

19 American mountaineer Annie Peck was born on this day in 1850. Beginning in 1895, she climbed mountains in Europe and the Americas. Her last climb was up Mount Madison in New Hampshire at age eighty-two.

20 On this day in 1968, former first lady Jacqueline Kennedy married Greek shipping tycoon Aristotle Onassis.

21 Ursula Le Guin, American author best known for her children's fantasy novels, was born on this day in 1929.

22 American Harriet Chalmers, the first foreign woman to travel to the inner regions of South and Central America, was born on this day in 1875. She founded the Society of Women Geographers and wrote regularly for *National Geographic* magazine.

23 On this day in 1934, American adventurer Jeanette Piccard set an altitude record for female balloonists when she ascended to 57,579 feet.

24 On this day in 1901, Annie Edson Taylor, a school-teacher from Michigan, became the first person to go over Niagara Falls in a barrel.

25 American country singer and comedienne Minnie Pearl was born on this day in 1912.

26 Hillary Rodham Clinton, the first first lady with an advanced degree (in law), and the first to become a U.S. senator (from New York, in 2000), was born on this day in 1947.

27 American poet Sylvia Plath was born on this day in 1932.

28 American actress Julia Roberts was born on this day in 1967.

29 On this day in 1966, the National Organization for Women (NOW) was founded.

30 On this day in 1944, dancer and choreographer Martha Graham debuted her ballet *Appalachian Spring* at the Library of Congress.

31 Juliette Gordon Low, founder of the Girl Scouts of the USA, was born on this day in 1860.

NOVEMBER

Gemstone: topaz
Flower: chrysanthemum

1 British writer Naomi Mitchison was born on this day in 1897.

2 Canadian country and rock singer k. d. lang was born on this day in 1961.

3 American actress and comedienne Roseanne Barr was born on this day in 1953.

4 First Lady Laura Welch Bush was born on this day in 1946.

5 On this day in 1968, Shirley Chisholm of New York was elected the first African American woman to serve in Congress. Her motto was "unbought and unbossed." She was a congresswoman from 1969 to 1983.

6 On this day in 1987, Tania Aebi returned to New York after sailing solo around the world for twenty-seven months. She was the first American woman and the youngest person ever to do so.

7 Canadian singer and songwriter Joni Mitchell was born on this day in 1943.

8 American actress Katharine Hepburn was born on this day in 1909.

9 Pulitzer Prize–winning American poet Anne Sexton was born on this day in 1928.

10 On this day in 1982, the Washington, D.C., Vietnam Veterans Memorial, designed by American architect Maya Lin, opened to visitors.

11 Abigail Adams, the politically influential first lady known for her widely published letters, was born on this day in 1744.

12 Elizabeth Cady Stanton, writer and founding mother of American feminism, was born on this day in 1815.

13 American actress and comedienne Whoopi Goldberg was born on this day in 1949.

14 Swedish author Astrid Lindgren, best known for creating the character Pippi Longstocking, was born on this day in 1907.

15 American painter Georgia O'Keeffe was born on this day in 1869.

16 American journalist Elizabeth Drew was born on this day in 1935.

17 On this day in 1637, religious leader Anne Hutchinson was banished from the Massachusetts Bay Colony after insisting that women had the right to express themselves on church matters.

18 Wilma Mankiller, American Indian activist who served as chief of the Cherokee Nation of Oklahoma, was born on this day in 1945.

19 American actress Jodie Foster was born on this day in 1963.

20 Pauli Murray, a civil rights lawyer and Episcopal priest — and the first African American to earn a doctorate from Yale Law School — was born on this day in 1910.

21 American actress Goldie Hawn was born on this day in 1945.

22 British novelist George Eliot, whose real name was Mary Ann Evans, was born on this day in 1819.

23 On this day in 1935, Ethel Leginska, the first woman to write and conduct an opera, presented her work *Gale* at the Chicago City Opera.

24 On this day in 1871, *Hit,* the memoir of Civil War surgeon and Congressional Medal of Honor winner Dr. Mary Edwards Walker, was published.

25 American singer Tina Turner was born on this day in 1942.

26 American abolitionist and feminist Sarah Grimké was born on this day in 1792.

27 British writer and actress Fanny Kemble was born on this day in 1809.

28 On this day in 1919, Lady Nancy Astor, an American socialite, became the first woman elected to the British parliament.

29 Louisa May Alcott, author of the classic American novel *Little Women,* was born on this day in 1832.

30 On this day in 1968, Diana Ross and the Supremes had a number one hit with "Love Child."

DECEMBER

Gemstone: turquoise
Flower: poinsettia

1 Today is Rosa Parks Day. On this day in 1955, Rosa Parks, a longtime activist with the National Association for the Advancement of Colored People (NAACP), refused to give up her seat on a bus in Montgomery, Alabama, so that a white man could have it. Parks was arrested, and the Montgomery black community launched a bus boycott that lasted for more than a year and became a turning point in the civil rights movement.

2 Tony Award–winning actress Julie Harris was born on this day in 1925.

3 On this day in 1990, Mary Robinson was sworn in as the first female president of Ireland.

4 On this day in 1978, Dianne Feinstein became the first female mayor of San Francisco.

5 Elizabeth Agassiz, educator and first president of Radcliffe College, was born on this day in 1822.

6 On this day in 1902, the first U.S. postage stamp picturing an American woman was issued. Martha Washington appeared on a lilac-colored eight-cent stamp.

7 Willa Cather, Pulitzer Prize–winning author of such novels as *My Àntonia* and *Death Comes for the Archbishop,* was born on this day in 1873.

8 Mary, Queen of Scots, was born on this day in 1542.

9 Grace Hopper, Navy rear admiral and computer pioneer, was born on this day in 1906.

10 On this day in 1869, Wyoming became the first state to grant women the right to vote in state and local elections.

11 Annie Jump Cannon, American astronomer who classified more than 350,000 stars, was born on this day in 1863.

12 Cathy Rigby, American Olympic gymnast and actress, was born on this day in 1952.

13 American civil rights activist Ella Baker was born on this day in 1903.

14 Margaret Chase Smith, the first woman elected to both houses of Congress, was born on this day in 1897. She was a U.S. representative from 1940 to 1948 and a U.S. senator from 1948 to 1972.

15 On this day in 1939, the movie *Gone With the Wind,* based on the Pulitzer Prize–winning novel by Margaret Mitchell, had its premiere in Atlanta, Georgia. Among the recipients of the movie's ten Academy Awards were Vivien Leigh for best actress and Hattie McDaniel for best supporting actress. McDaniel was the first African American to win an Academy Award.

16 Margaret Mead, an American anthropologist who wrote more than thirty books, was born on this day in 1901.

17 Deborah Sampson, a soldier in the Revolutionary War, was born on this day in 1760. In order to fight, she disguised herself as a man and went by the name Robert Shurtleff.

18 Gladys Henry Dick, a microbiologist and physician who found the bacterial cause of scarlet fever, was born on this day in 1881.

19 French singer Edith Piaf was born on this day in 1915.

20 On this day in 1928, the first Broadway theater named for a living actress, the Ethel Barrymore Theater, opened its doors.

21 On this day in 1898, French-Polish chemist Marie Curie and her husband, Pierre Curie, discovered radium, a radioactive element.

22 American journalist and news anchor Diane Sawyer was born on this day in 1945.

23 Madame C. J. Walker, businesswoman and manufacturer of beauty products who became the first African American millionaire, was born on this day in 1867.

24 Elizabeth Chandler, American author and abolitionist, was born on this day in 1807.

25 Annie Lennox, Scottish lead singer for the band Eurythmics and later a successful solo artist, was born on this day in 1954.

26 Susan Butcher, four-time winner of the Iditarod dogsled race, was born on this day in 1956.

27 On this day in 1920, the curtain went up on the Pulitzer Prize–winning play *Miss Lulu Bett,* the first Broadway play written by a woman, Zona Gale.

28 Carol Ryrie Brink, American author of *Caddie Woodlawn,* was born on this day in 1895.

29 Mary Tyler Moore, American actress and star of the television series *The Mary Tyler Moore Show,* was born on this day in 1936.

30 Tracey Ullman, British actress and comedienne, was born on this day in 1959.

31 On this day in 1993, American singer Barbra Streisand returned to the stage for her first paid concert in twenty-two years.

American History

LANDMARKS

At least 12,000 years ago According to a theory accepted by most anthropologists, the first women arrive in North America via the Bering land bridge from Asia.

At least 2,000 years ago Women play important roles in the hundreds of different American Indian cultures that thrive before European arrival in the 1500s.

Women's roles are as varied as their societies. Some women gather food or plant crops; some make tools and build houses; some participate in trade. In some societies, community life and economics are organized around female kinship. In many, older women are important leaders; they might choose the chief, arrange marriages, or run the treasury.

1587 Virginia Dare is the first child born in America to English parents (Roanoke Island, Virginia).

1848 The first women's rights convention is held in Seneca Falls, New York. After two days of discussion and debate, sixty-eight women and thirty-two men sign a Declaration of Sentiments, which sets the agenda for the women's rights movement. A set of resolutions calls for equal treatment of women and men under the law and voting rights for women.

1850 The first National Women's Rights Convention takes place in Worcester, Massachusetts, attracting more than one thousand participants. National conventions are held yearly (except in 1857) through 1860.

1861–1865 The Civil War. An estimated 3,200 women

served as volunteer nurses for the Northern and Southern armies.

1874 The Woman's Christian Temperance Union (WCTU) is founded to improve the morality of the nation, in particular by protesting alcoholic beverages.

1896 The National Association of Colored Women is formed, bringing together more than one hundred black women's clubs.

1917–1918 U.S. involvement in World War I. About ten thousand American women serve as nurses for the military.

1919 The federal Woman Suffrage Amendment, originally written by Susan B. Anthony and introduced in Congress in 1878, is passed by both houses of Congress.

1920 The Nineteenth Amendment to the U.S. Constitution, granting women the right to vote, is signed into law.

1935 Mary McLeod Bethune organizes the National Council of Negro Women, a coalition of black women's groups that lobbies against job discrimination, racism, and sexism.

1941–1945 U.S. involvement in World War II. About one hundred thousand women serve as WACs (members of the Women's Army Corps), and about eighty-six thousand as WAVES (Women Appointed for Voluntary Emergency Service). On the home front, more than six million women fill industrial jobs to help the war effort.

1963 Betty Friedan publishes her highly influential book *The Feminine Mystique,* which describes the dissatisfaction felt by many American housewives. The book becomes a bestseller and helps to launch the modern women's rights movement.

DID YOU KNOW?

Some Cowgirls Already Had the Vote

In 1920, American women won the right to vote in national elections — one of the most significant achievements in their history. But did you know that women had already won the right to vote in a number of states? Actually, the territory of Wyoming was the first place in the United States to pass a woman suffrage law, in 1869. Women began serving on juries in the territory the following year.

In 1893, Colorado became the first state to adopt an amendment granting women the right to vote. Utah and Idaho followed suit in 1896; Washington State in 1910; California in 1911; Oregon, Kansas, and Arizona in 1912; Alaska and Illinois in 1913; Montana and Nevada in 1914; New York in 1917; and Michigan, South Dakota, and Oklahoma in 1918.

1966 The National Organization for Women (NOW) is founded by a group of feminists including Betty Friedan. The largest women's rights group in the nation, NOW seeks to end sexual discrimination, especially in the workplace, by means of legislative lobbying, litigation, and public demonstrations.

1972 The Equal Rights Amendment (ERA) is passed by Congress and sent to the states for ratification. Originally drafted in 1923, the amendment reads: "Equality of rights under the law shall not be denied or abridged by the United States or by any State on account of sex." The amendment dies in 1982 when it fails to achieve ratification by a minimum of thirty-eight states.

WOMEN'S HISTORY MONTH

Women's History Month has been celebrated nationally since 1987 (and as Women's History Week from 1981–1986). There's a lot to celebrate!

In the nineteenth century, when the women's rights movement was born, women were second-class citizens. They were just beginning to gain admission to colleges. They were prohibited from entering many professions. Married women had to surrender most of their rights, including the right to own property, to their husbands. Women could not participate in national elections. But women worked to improve their status. Women's History Month celebrates those trailblazers who helped women to secure a more equal place in society.

INTERNATIONAL WOMEN'S DAY

International Women's Day is celebrated on March 8 to honor the women of the world. The holiday was first celebrated in the United States on February 28, 1909, under the leadership of the Socialist Party of America. On the eve of World War I it became an annual event that was part of the peace movement in Europe. (In Russia today it is a major holiday — celebrated, like Mother's Day, with flowers or breakfast in bed — on which men show appreciation for the women in their lives.)

Since 1975, the United Nations has sponsored International Women's Day, and its date has been fixed at March 8. The UN calls it a day to honor "ordinary women as makers of history."

RELIGION

Even though women have only recently been permitted to hold official roles in many religions, they have always been central to American religious life. Unofficially, women have often been the primary carriers and creators of religious culture.

Religion has also been an arena for American female activists. Many abolitionists and other early social reformers were motivated in part by religious belief. Beginning in the 1800s, numerous Christian women, black and white, and Jewish women founded religious schools and aid organizations. Many of the African American women who helped power the civil rights movement in the 1960s drew strength from their religions and organized through their churches.

Here are some notables in the history of women and American religion.

First community of nuns in the thirteen colonies: A Carmelite convent near Port Tobacco, Maryland, established by Mother Bernardina Matthews in 1790.

First female minister in a recognized U.S. denomination: Antoinette (Brown) Blackwell, in 1853. She was ordained in the Congregational Church but later became a Unitarian.

First major religion founded by an American woman: The Church of Christ, Scientist, established by Mary Baker Eddy in 1879.

First U.S. citizen to become a saint: Mother Frances Xavier Cabrini (1850–1917), in 1946. She was born in Italy.

First native-born American to become a saint: Elizabeth Ann Seton, in 1975. She had established the first American community of the Sisters of Charity, in Emmitsburg, Maryland, in 1809.

First female American rabbi: Sally Jean Priesand, in 1972. She was ordained at Hebrew Union College in Cincinnati.

First female Episcopal bishop in the United States: Barbara Harris, in 1989. She was also among the first African American women ordained as Episcopal priests.

Largest religious women's organization in the United States: The Relief Society (Church of Jesus Christ of Latter-day Saints), founded in 1842. It is also one of the largest women's organizations in the world.

WOMEN WHO LEFT THEIR STAMPS ON HISTORY

Each of the following women has made a significant contribution to American history. As a result, they (or, in a few cases, works of art created by them) have been pictured on U.S. postage stamps.

Name	Year Issued	Contribution
Queen Isabella of Spain	1893	Funded Columbus's trips to the New World
Martha Washington	1902	First U.S. first lady
Pocahontas	1907	Indian princess and peacemaker

Name	Year Issued	Contribution
Molly Pitcher	1928	Participated in the 1778 Battle of Monmouth
Eleanor Roosevelt	1930 1984 1998	Diplomat, social reformer, and first lady
Susan B. Anthony	1936 1955	Leader for women's rights
Virginia Dare	1937	First European child born on American soil, in 1587
Louisa May Alcott	1940	Author of *Little Women*
Frances E. Willard	1940	Educator and suffragist
Jane Addams	1940	Founded Chicago's Hull House, a social welfare center
Clara Barton	1948	Founded the American Red Cross
Juliette Gordon Low	1948	Founded Girl Scouts of the USA
Moina Michael	1948	Initiated the Veterans of Foreign Wars fundraising drive in 1915
Betsy Ross	1952	The nation's most famous flag maker
Sacajawea	1954	Shoshone guide who led the Lewis and Clark expedition of 1804
Amelia Earhart	1963	First woman to fly solo and nonstop across the Atlantic
Mary Cassatt	1966 1988	Impressionist painter
Lucy Stone	1968	Abolitionist and women's rights leader
Grandma Moses	1969	Folk art painter
Emily Dickinson	1971	Author of more than 1,700 poems
Willa Cather	1973	Pulitzer Prize–winning novelist
Elizabeth Blackwell	1973	First American female doctor

Name	Year Issued	Contribution
Sybil Ludington	1975	Sixteen-year-old Revolutionary War heroine
Clara Maass	1976	U.S. Army nurse who helped advance the fight against yellow fever
Harriet Tubman	1978	Risked her life to help bring hundreds of slaves to freedom
Emily Bissell	1980	Leader in the fight against tuberculosis
Helen Keller and Anne Sullivan	1980	Famous blind and deaf student and her extraordinary teacher
Dolley Madison	1980	First lady known as an outstanding hostess
Frances Perkins	1980	First female member of a presidential cabinet, under F. D. Roosevelt
Edith Wharton	1980	Pulitzer Prize–winning novelist
Rachel Carson	1981	Environmentalist and author of the landmark book *Silent Spring*
Edna St. Vincent Millay	1981	Pulitzer Prize–winning poet
Mildred "Babe" Didrikson Zaharias	1981	Star in track, golf, baseball, and basketball
Mary Walker	1982	Doctor who cared for the sick and wounded during the Civil War
Dorothea Dix	1983	Nineteenth-century social reformer
Pearl S. Buck	1983	Nobel and Pulitzer Prize–winning novelist
Lillian M. Gilbreth	1984	Early twentieth-century engineer
Abigail Adams	1985	First lady and influential letter writer

Name	Year Issued	Contribution
Mary McLeod Bethune	1985	Presidential adviser and founder of Bethune-Cookman College
Belva Ann Lockwood	1986	Presidential candidate in 1884 and 1888
Margaret Mitchell	1986	Pulitzer Prize–winning novelist
Sojourner Truth	1986	First black woman to speak publicly against slavery, in the 1840s
Julia Ward Howe	1987	Composer of "The Battle Hymn of the Republic"
Mary Lyon	1987	Founder of Mount Holyoke College
Helene Madison	1990	Olympic gold medalist in 1932 in swimming
Marianne Moore	1990	Pulitzer Prize–winning poet
Ida B. Wells	1990	Antilynching activist
Hazel Wightman	1990	Early nineteenth-century tennis champion
Fanny Brice	1991	Vaudeville singer and comedienne
Harriet Quimby	1991	First woman to fly across the English Channel, in 1912
Dorothy Parker	1992	Poet and short story writer
Patsy Cline	1993	Country singer
Grace Kelly	1993	Film actress and princess
Dinah Washington	1993	Blues singer
Clara Bow, ZaSu Pitts, Theda Bara	1994	Silent film actresses
Nellie Cashman	1994	Social reformer
Ethel Waters, Bessie Smith, Billie Holiday, Mildred Bailey, Ethel Merman	1994	American singers
Annie Oakley	1994	Sharpshooter

Name	Year Issued	Contribution
Virginia Apgar	1994	Doctor who developed a newborn-assessment method
Ruth Benedict	1995	Anthropologist
Mary Chesnut, Phoebe Pember	1995	Heroic Confederate women
Bessie Coleman	1995	First woman to earn an international pilot's license
Alice Hamilton	1995	Pioneer in industrial medicine
Marilyn Monroe	1995	Actress
Alice Paul	1995	Author of the Equal Rights Amendment
Jacqueline Cochran	1996	First female pilot to break the sound barrier
Georgia O'Keeffe	1996	Abstract painter
Dorothy Fields	1997	Popular songwriter of the 1920s and 1930s
Lily Pons, Rosa Ponselle	1997	Opera singers
Mary Breckinridge	1998	Founder of the Frontier Nursing Service
Mahalia Jackson, Roberta Martin, Sister Rosetta Tharpe, Clara Ward	1998	Gospel singers
Margaret Mead	1998	Anthropologist
Madame C. J. Walker	1998	Businesswoman
Ayn Rand	1999	Novelist
Patricia Roberts Harris	2000	Lawyer and political adviser
Louise Nevelson	2000	Sculptor
Hattie Caraway	2001	First woman elected to the U.S. Senate
Rose O'Neill	2001	Illustrator
Lucille Ball	2001	Comedienne and actress
Frida Kahlo	2001	Mexican artist
Nellie Bly, Marguerite Higgins, Ethel Payne, Ida Tarbell	2002	Journalists

FIVE WOMEN TO KNOW

Sacajawea (1784?–1884?) was a Shoshone Indian who acted as a geographic guide, diplomat, and interpreter for the Lewis and Clark expedition from 1804 to 1806, leading the explorers from the Great Plains to the Pacific coast. It is believed that they could not have survived without her expertise. At the time, Sacajawea was just a teenager. In 2000 the U.S. Mint issued a circulating dollar coin with her likeness.

Sojourner Truth (1797?–1883) was the most prominent African American woman in both the abolitionist movement and the early feminist movement. Born a slave called Isabella, she took the name Sojourner Truth some years after achieving freedom and traveled as a speaker. She was known for her charisma, her majestic height, and her rousing "Ain't I a Woman?" speech at a women's rights convention in 1851.

Susan B. Anthony (1820–1906) was a leader in the fight for women's rights for nearly seventy years. She began organizing for equal pay as a teenage schoolteacher and in the 1850s became a national leader for women's suffrage. She lectured widely in the United States and Europe and wrote a three-volume history of the suffrage movement. The U.S. Mint issued a circulating dollar coin with her image in 1979.

Rosa Parks (b. 1913), an activist with the Montgomery, Alabama, National Association for the Advancement of Colored People (NAACP), made history in 1955 when she refused to give up her seat on a bus for a white passenger. She was arrested, and the Montgomery black community launched a bus boycott that became one of the pivotal events in the civil rights movement.

"The Pledge of Allegiance says 'Liberty and Justice for All.' Which part of 'All' don't you understand?"

— Pat Schroeder, former U.S. representative

"Never doubt that a small group of thoughtful committed people can change the world: indeed it's the only thing that ever has!"

— Margaret Mead, American anthropologist

"We hold these truths to be self-evident, that all men and women are created equal . . ."

— Elizabeth Cady Stanton, feminist, in the *Declaration of Sentiments* (1848)

"The test of a civilization is in the way that it cares for its helpless members."

— Pearl S. Buck, American writer, in *My Several Worlds* (1954)

"I never doubted that equal rights was the right direction. Most reforms, most problems are complicated. But to me there is nothing complicated about ordinary equality."

— Alice Paul, American lawyer and feminist

Gloria Steinem (b. 1934) has been an influential journalist and speaker for more than forty years. In 1972, she was a founder and editor of *Ms.* magazine, which became a major forum for women's voices and helped launch Steinem as an icon of the modern feminist movement. Her collection of essays *Outrageous Acts and Everyday Rebellions* (1983) is considered a classic.

American Places

THE BIRTHPLACE OF
WOMEN'S RIGHTS

Some people consider the single most important place in U.S. women's history to be Seneca Falls, New York, where on July 19, 1848, the first women's rights convention was held. Seneca Falls was the home of Elizabeth Cady Stanton, who, along with abolitionist Lucretia Mott, organized the convention.

The National Women's Hall of Fame was founded in 1969 in Seneca Falls. The hall inducts distinguished women and offers programs and exhibits in Seneca Falls; New York's Finger Lakes area; Washington, D.C.; and elsewhere. The mission of the National Women's Hall of Fame is "to honor in perpetuity those women, citizens of the United States of America, whose contributions to the arts, athletics, business, education, government, the humanities, philanthropy and science, have been the greatest value. . . ."

Since 1980, Seneca Falls has also been the site of the Women's Rights National Historic Park. Among its numerous features are the chapel where the 1848 convention was held; a one-hundred-foot-long wall engraved with the Declaration of Sentiments, an equal rights proclamation that was signed at the convention; and the house where Stanton and her family lived after moving from Boston in 1847. Stanton said that the difficulty of daily life in Seneca Falls, where for the first time she was alone in caring for a large house and three small children, made her a feminist.

TEN MORE PLACES WHERE
WOMEN MADE HISTORY

The museums of the Seneca Falls area could keep you busy — and inspired — for days. But there are many other places where you can learn more about the courageous women who made a difference in U.S. history. Here are some that are open to visitors.

Emily Dickinson House
Amherst, Massachusetts
Poet Emily Dickinson (1830–1886) was born in this house and lived most of her life there, rarely leaving its beautiful grounds. After Dickinson's death, her sister Lavinia discovered hundreds of poems hidden in the house. Thanks to Lavinia, these poems were published, and Dickinson is now regarded as one of the greatest American poets.

Harriet Tubman Home for the Aged
Auburn, New York
For more than a decade the legendary Harriet Tubman (1820–1913) risked her life to lead hundreds of enslaved people out of the South to freedom in the North. Tubman also worked as a nurse and scout during the Civil War. Her home for the aged was her last major project. It is now a museum honoring Tubman's life and work.

Iolani Palace
Honolulu, Hawaii
The only royal palace in the United States, Iolani was the official residence of the Hawaiian monarchy from 1882 until foreign merchants overthrew Queen Liliuokalani in 1893. The queen was also imprisoned there for eight months in 1895 following efforts to restore her to the

throne. Portraits of several Hawaiian queens are on display at Iolani.

Jane Addams Hull-House Museum
Chicago, Illinois
Founded by Jane Addams in 1889, Chicago's Hull House may be the most famous of the "settlement houses" that improved life for immigrant families in poor city neighborhoods. Hull House offered classes, day care, job assistance, and a place for labor unions to organize. Like most other settlement houses, Hull House was staffed mainly by women.

Juanita Craft House
Dallas, Texas
Both Martin Luther King Jr. and President Lyndon Johnson visited this ordinary home to seek advice from the extraordinary civil rights activist Juanita Craft (1902–1985). Craft, a dressmaker, joined the National Association for the Advancement of Colored People (NAACP) in 1935. By 1958, she had founded 182 rural chapters of the NAACP.

Laura Ingalls Wilder Memorial Society
De Smet, South Dakota
Six of Laura Ingalls Wilder's Little House books are set in the small prairie town of De Smet. Today you can tour her family's first Dakota house, their final home in town, and a replica of Laura's school. You can also visit the Big Slough, the five cottonwood trees Laura's Pa planted to honor "his girls," and the cemetery where the family is buried.

Lowell National Historical Park
Lowell, Massachusetts
In the 1830s, the mills in the booming industrial town of Lowell employed women almost exclusively. Thousands of young women left their farm homes to live together at Lowell boardinghouses and work in the mills. Today, tours

and exhibits at Lowell National Historic Park bring to life these women's contributions to the Industrial Revolution.

Mary McLeod Bethune Council House
Washington, D.C.
This three-story townhouse was the Washington, D.C., residence of educator Mary McLeod Bethune (1875–1955) and the original headquarters of the National Council of Negro Women. It is now a National Historic Site that hosts a variety of programs. A carriage house in back contains the National Archives for Black Women's History.

Orchard House
Concord, Massachusetts
Author Louisa May Alcott (1832–1888) lived in Orchard House with her family for nearly twenty years and is thought to have written her classic novel *Little Women* there. Now a National Historic Landmark, Orchard House still looks much as it did when the Alcotts lived there. Visitors can tour the home and imagine how the family lived.

Susan B. Anthony House
Rochester, New York
A National Historic Landmark, this house was the home and headquarters of feminist Susan B. Anthony (1820–1906) for nearly half a century. Countless meetings to organize for women's rights were held in this house. In 1872, Anthony was famously arrested in the house because she had voted in the presidential election — a criminal act for a woman.

STATE HALLS OF FAME

A number of states have halls of fame that honor the
renowned women who once called those states home or
helped to make them great. These are a few to know.

Alabama Women's Hall of Fame
Marion, Alabama
Founded 1970
Honorees include: writer and educator Helen Keller, actress
Tallulah Bankhead, novelist and painter Zelda Fitzgerald,
folklorist Ruby Pickens Tartt

Arizona Women's Hall of Fame
Phoenix, Arizona
Founded 1979
Honorees include: Yavapai Prescott Indian chief Viola
Jimulla, women's health activist Margaret Sanger, rancher
Amy Cornwall Neal, historian and poet Sharlot M. Hall

Colorado Women's Hall of Fame
Denver, Colorado
Founded 1985
Honorees include: *Titanic* heroine Molly Brown, former
Israeli prime minister Golda Meir, politician Pat Schroeder,
author and activist Mildred Pitts Walter

Connecticut Women's Hall of Fame
Hartford, Connecticut
Founded 1994
Honorees include: opera singer Marian Anderson, aboli-
tionist Prudence Crandall, actress Katharine Hepburn,
writer Madeleine L'Engle

Florida Women's Hall of Fame
Tallahassee, Florida
Founded 1982
Honorees include: writer Zora Neale Hurston, educator
Mary McLeod Bethune, athlete Althea Gibson, former U.S.
attorney general Janet Reno

Iowa Women's Hall of Fame
Des Moines, Iowa
Founded 1975
Honorees include: feminist Amelia Jenks Bloomer, journalist
Mildred Wirt Benson, suffrage movement leader Carrie
Chapman Catt, first female American lawyer Arabella
Mansfield

Maine Women's Hall of Fame
Augusta, Maine
Founded 1990
Honorees include: former U.S. senator Margaret Chase
Smith, scientist Elizabeth S. Russell, U.S. senator Olympia
Snowe

Maryland Women's Hall of Fame
Annapolis, Maryland
Founded 1985
Honorees include: nurse Clara Barton, fashion designer
Claire McCardell, abolitionist Harriet Tubman, ecologist
Rachel Carson

Michigan Women's Historical Center and Hall of Fame
Lansing, Michigan
Founded 1987
Honorees include: actress Ellen Burstyn, former first lady
Betty Ford, singer Aretha Franklin, civil rights activist Rosa
Parks, comedienne Lily Tomlin

Ohio Women's Hall of Fame
Columbus, Ohio, and regional exhibits
Founded 1978
Honorees include: photographer Berenice Abbott, sharp-shooter Annie Oakley, poet Nikki Giovanni, writer Toni Morrison

Texas Women's Hall of Fame
Denton, Texas
Founded 1984
Honorees include: astronaut Mae Jemison, former first lady Barbara Bush, businesswoman Mary Kay Ash, former Texas governor Ann Richards

FIVE WOMEN'S MUSEUMS TO KNOW

Since 1975, the **National Cowgirl Museum and Hall of Fame,** in Fort Worth, Texas, has honored cowgirls and other heroines of the American West. The museum features exhibits, lectures, and other events. Potter Maria Martinez, writer Willa Cather, and Indian guide Sacajawea are among the women honored in the hall of fame.

The **National Museum of Women in the Arts,** in Washington, D.C., is the only museum in the world that focuses on the achievements of female artists. The museum's permanent collection includes more than three thousand works of art, from sculpture to paintings to photography, and covers the sixteenth to the twenty-first century.

Founded in 1991, the **Women of the West Museum,** in Denver, Colorado, offers traveling exhibits, online activities, and a variety of educational programs. Topics explored in recent exhibits include the early fight for women's suffrage in the West and what life was like for the many female homesteaders who lived in sod houses.

The **International Women's Air and Space Museum**, in Dayton, Ohio, was established in 1986 to honor the history of women who took flight in our atmosphere and beyond. The museum covers adventurer pilots such as Amelia Earhart, female pilots in the military, and those who ventured into the newer frontier — space.

The **U.S. Army Women's Museum**, in Fort Lee, Virginia, opened in May 2001. The museum is dedicated to preserving the history of women who served in the army from the Revolutionary War through today. It is located on the former site of the Women's Army Corps Center and School, where thousands of army women were trained.

WOMEN ON PEDESTALS

These are monumental women! Because of their achievements, their likenesses have been carved in stone for all to see and remember.

Who: Alice Cogswell, first student at the first school for the deaf in the United States.
What: A bronze statue of young Alice shows her signing the letter A with her teacher, Thomas Hopkins Gallaudet.
Where: Gallaudet University, Washington, D.C.

Who: Amelia Earhart, pioneer aviator, called "the Golden Girl of Aviation"
What: A seven-foot-tall statue covered with gold leaf, with airplane propellers embedded in the base

Where: North Hollywood, California

Who: Sybil Ludington, sixteen-year-old Revolutionary War
hero
What: A bronze statue showing her on horseback during
her nighttime ride to warn soldiers
Where: Carmel, New York

Who: Annie Moore, fifteen-year-old from Ireland who was
the first immigrant to pass through the receiving room
at Ellis Island when it opened in 1892
What: A bronze statue showing Annie with a satchel in her
hand and a hopeful expression on her face
Where: Ellis Island, New York Harbor, New York

Who: Esther Morris, who helped make Wyoming the first
state to grant women the right to vote
What: A shiny brass statue showing her as a young woman
carrying flowers and a portfolio
Where: Entrance to the state capitol building, Cheyenne,
Wyoming

Who: Annie Oakley (Phoebe Anne Mozee), famous Wild
West sharpshooter
What: A life-size bronze statue showing her standing,
holding her rifle by her side
Where: Greenville, Ohio

Who: Pocahontas, a Powhatan Indian who at the age of
ten helped the Jamestown colonists and saved the life of
their leader, Captain John Smith
What: A life-size outdoor statue showing her with open,
helping arms
Where: Jamestown, Virginia

Who: Eleanor Roosevelt, first lady, humanitarian
What: An eight-foot bronze statue of her as an older
woman leaning against a rock
Where: Riverside Park, New York City

Who: Sacajawea, guide and scout for the Lewis and Clark expedition
What: A twelve-foot bronze statue showing Sacajawea with her baby strapped to her back
Where: Bismarck, North Dakota

Who: Gertrude Stein, author
What: A bronze statue showing her seated, looking, as she once described herself, "like a great Jewish Buddha"
Where: Bryant Park, New York City

Who: Phillis Wheatley, famous slave poet
What: A bronze statue by Elizabeth Catlett, dedicated by black female poets
Where: Jackson, Mississippi

Who: Women of the Vietnam era
What: The Vietnam Women's Memorial, a sculpture portraying one woman caring for a wounded soldier, another kneeling with her head bowed, and a third searching the sky for airborne help.
Where: Washington, D.C.

GIRLS ACROSS AMERICA

From Tracy, California, to Chelsea, Vermont, the map of the United States is dotted with girls' names. Here are a few.

Allison, Iowa
Amy, Kansas
Angela, Montana
Beatrice, Nebraska
Bethany, Missouri
Beverly, Massachusetts
Camilla, Georgia
Charlotte, North Carolina
Chelsea, Vermont
Clare, Michigan
Dana, Indiana
Donna, Texas
Dorothy, Minnesota
Eileen, Illinois
Elaine, Arkansas
Elizabeth, New Jersey
Faith, South Dakota
Florence, Kentucky
Gail, Texas
Greer, South Carolina
Hannah, North Dakota
Helen, Maryland
Henrietta, New York
Ida, Iowa
Irene, South Dakota
Isabel, Kansas
Jasmine, Arkansas
Jeannette, Pennsylvania
Kelly, Kansas
Kimberly, Idaho
Laurel, Delaware

Louisa, Virginia
Magdalena, New Mexico
Meredith, New Hampshire
Mona, Utah
Nadine, New Mexico
Natalie, Pennsylvania
Neenah, Wisconsin
Olivia, Minnesota
Olympia, Washington
Paola, Kansas
Patricia, Texas
Pauline, South Carolina
Rachel, West Virginia
Ramona, California
Rebecca, Georgia
Shannon, Mississippi
Sharon, Pennsylvania
Susan, Virginia
Tallulah, Louisiana
Theresa, New York
Tracy, California
Una, South Carolina
Ursula, Arkansas
Victoria, Texas
Virginia, Illinois
Wanda, Minnesota
Winona, Mississippi
Xenia, Ohio
Yolande, Alabama
Zelda, Kentucky
Zena, New York

Books

LANDMARKS

1650 Anne Bradstreet's first book of poems, *The Tenth Muse Lately Sprung Up in America,* is published in England, making her the first published American female writer.

1766 Mary Katherine Goddard and her widowed mother become publishers of the *Providence Gazette* newspaper and the annual *West's Almanack,* making them the first female publishers in America. In 1789, Goddard opens a Baltimore bookstore; she is probably the first woman in America to own a bookshop.

1791 Actress and schoolteacher Susanna Haswell Rowson's melodramatic novel *Charlotte: A Tale of Truth* is published. Later known as *Charlotte Temple,* it becomes a bestseller with more than 150 printings.

1826 Lydia Maria Child serves as editor of the *Juvenile Miscellany,* the first children's magazine in the United States.

1851 Harriet Beecher Stowe's first novel about slavery, *Uncle Tom's Cabin,* is featured as a serial in an abolitionist newspaper. Published as a book the following year, it becomes an international bestseller and has a profound impact on public opinion about slavery.

1859 Harriet Wilson becomes the first African American novelist with the publication of *Our Nig; or, Sketches from the Life of a Free Black.*

1912 The influential editor Harriet Monroe founds *Poetry: A Magazine of Verse,* which introduces American readers to such poets as Carl Sandburg and Robert Frost.

1920s In New York, African American women such as Zora Neale Hurston, Helene Johnson, and Nella Larsen become literary figures in the period of artistic flourishing that will come to be known as the Harlem Renaissance.

1921 Working in a Greenwich Village basement, with the noise of a speakeasy overhead, Lila Acheson Wallace and her husband found *Reader's Digest.* She serves as editor, manager, and art buyer for the magazine.

1927 Christine Quintasket, also known as Mourning Dove, publishes what is probably the first novel by an American Indian woman: *Cogewea, the Half Blood: A Depiction of the Great Montana Cattle Range.* It incorporates her observations of turn-of-the-century life on the Great Plains, including the last roundup of wild bison.

1950s The vivid and realistic short stories of Flannery O'Connor and Eudora Welty bring female voices to the fertile tradition of Southern literature.

1971 *Ms.* magazine is first published as a sample insert in *New York* magazine; three hundred thousand copies sell out in eight days. The first regular issue is published in July 1972.

1992–1993 Mona Van Duyn is the first woman to serve as poet laureate of the United States. (The national poet laureate is appointed by the Library of Congress each year and works to increase the public's appreciation of poetry.)

1993 Toni Morrison becomes the first American woman to win the Nobel Prize for literature.

1993–1995 Pulitzer Prize–winning poet Rita Dove is named poet laureate of the United States. She is the youngest person and the first African American to hold the position.

TEN GREAT GIRL CHARACTERS: NOVELS

Alice
In *Alice's Adventures in Wonderland,* by Lewis Carroll
Alice, an impressionable Victorian girl of seven and a half, falls down a rabbit hole into Wonderland, where she has many strange and curious adventures. Alice is one of the earliest classic child characters in literature.

Amelia Bedelia
In the Amelia Bedelia series, by Peggy Parish
Amelia Bedelia is a maid who interprets her duties literally. When she is asked to *dust* the furniture, she douses it with talcum powder! To *change* the towels, she cuts holes in them! Her confusion makes us see how many ways there are to interpret ordinary statements.

Anastasia Krupnik
In the Anastasia Krupnik series, by Lois Lowry
Whether she's stirring up trouble or using her imagination and sense of humor to get out of yet another predicament, Anastasia is refreshingly real. It's easy to relate to her challenges at school, with family, and in an ever changing world.

Annabel Andrews
In *Freaky Friday* and *A Billion for Boris,* by Mary Rodgers
Cranky teenager Annabel Andrews gets herself into the strangest situations. When we first meet her, she has switched bodies with her mother. Later, she begins to organize her life around a television that shows tomorrow's news.

Claudia Kincaid
In *From the Mixed-Up Files of Mrs. Basil E. Frankweiler,*
by E. L. Konigsburg
When Claudia Kincaid is sick of being told what to do at
home, she runs away to live in the Metropolitan Museum
of Art in New York City. Her brother Jamie joins her on an
adventure that leads them to the home of Mrs. Basil E.
Frankweiler, where Claudia unravels a mystery.

Harriet Welsch
In *Harriet the Spy,* by Louise Fitzhugh
Harriet M. Welsch, age eleven, wants to see and know
everything in her quest to become a famous writer. Notebook in hand, she spies on her friends, trying to trap them
in wrongdoings. But what happens when her notes acci-
dentally fall into her friends' hands?

Hermione Granger
In the Harry Potter series, by J. K. Rowling
The very clever Hermione Granger is probably the most
talented student in her year at Hogwarts School of
Witchcraft and Wizardry — and, some days, probably the
most annoying. But she's also a brave adventurer and a
loyal friend.

Laura
In the Little House series, by Laura Ingalls Wilder
The central character is really the author, who started to
write about her childhood in 1932, when she was sixty-
three years old. These memoirs tell the story of Laura, a girl
who moves with her family from a log cabin in Wisconsin
across the prairie states. The books recall her life from her
young tomboy days to her adulthood.

Ramona Quimby
In the Ramona series, by Beverly Cleary
Ramona Geraldine Quimby lives in an American town with
her parents and her older sister, Beezus. Ramona is a lively,

naughty girl whose parents think she is adorable and for-
give all her inventive, crazy ideas (which usually misfire).

Rose Rita Pottinger
In *The Ghost in the Mirror* and other books, by John Bellairs
Rose Rita Pottinger has an outdoorsy spirit, a gift for tall
tales, and a keen interest in magic. Fortunately, she also
has a cool head when confronting evil sorcerers or going
on impromptu time-travel journeys.

FIVE GREAT FEMALE
CHARACTERS: COMIC BOOKS

There's a whole world of comic books beyond *Superman*
and *Peanuts.* In recent years, more women have been
working in this literary form, and not coincidentally, there
are more interesting female characters than ever before.
Here are five of these heroines — some brave, some funny,
and all very original — who take center stage in a new era
of graphic novels.

Lady Jain
In the Castle Waiting series, by Linda Medley
This fairy-tale series tells what happens *after* "happily ever
after" comes to an end. Lady Jain leaves her Prince Not-So-
Charming for a journey of discovery. She is aided by
witches, monsters, and other surprising characters, including an order of bearded nuns.

Pastil
In the Pastil series, by Francesca Ghermandi
This surreal series from Italy stars Pastil, a little girl with a
pill-shaped head, in whose woody world lurks an array of
mysterious dangers. Ghermandi's highly imaginative soft-

DID YOU KNOW?

Mystery Solvers

Girl detectives were some of the earliest stars of popular series fiction for girls — and some of the earliest role models for modern independent women. The dramatic exploits of amateur detectives Nancy Drew and Judy Bolton and career girl sleuth Beverly Gray captivated armchair detectives in the 1920s. During the 1940s, readers were introduced to the mystery-solving skills of nurse detective Cherry Ames, flight attendant Vicki Barr, and teenage tomboy Trixie Belden. The 1950s brought reporter Sally Baxter and advertising agency employee Connie Blair onto the case.

The great age of the girl detective series has come and gone, though mystery is a part of many contemporary novels, such as those by Vivien Alcock and Natalie Babbitt. But Nancy Drew is still at the top of her game, more than seventy years after her debut. Where mystery lurks, Nancy can still be found sleuthing.

pencil illustrations have international appeal — the series has no words.

Rose
In *Rose,* by Jeff Smith and Charles Vess
The medieval adventures of the teenage princess Rose include triumphant battles with not only dragons but also Giant Hairy Rat Creatures. This book is a beautifully illustrated prequel to the *Bone* series, in which Rose appears as the aged warrior Gran'ma Ben.

Scary Godmother

In the Scary Godmother series, by Jill Thompson

Scary Godmother isn't all that scary, even though she lives on the Fright Side (just to the left of our universe) and shares her house with a skeleton in the closet and a monster under the bed. She's a humorous and helpful godmother to a little girl named Hannah Marie.

Sephie

In the Meridian series, by Barbara Kesel and others

The teenage Sephie was born to rule Meridian, a sky-city floating high above Earth, where ships sail on the wind. But a strange mark on her forehead threatens to change her destiny — and puts her beloved sky-city at risk.

MYSTERY INITIALS

It used to be common for authors to use only their first initials and last names — it was a touch of formality that was considered suitable behavior in the lofty profession of writer. But some female authors have used their initials in order to hide their gender. For example, Joanne Kathleen Rowling, author of the Harry Potter books, decided to call herself "J. K." on the advice of her publisher, so that her books would have greater appeal to boys. Until Rowling became a celebrity, many readers assumed that Harry's creator was a man. Here are some other authors you might not have known were women.

L. M. Boston

Known for: Green Knowe series
Full name: Lucy Maria Boston

S. E. Hinton
Known for: *The Outsiders*
Full name: Susan Eloise Hinton

E. L. Konigsburg
Known for: *From the Mixed-Up Files of Mrs. Basil E. Frankweiler*
Full name: Elaine Loeb Konigsburg

L. M. Montgomery
Known for: *Anne of Green Gables*
Full name: Lucy Maud Montgomery

☞ SHE SAID IT ☜

"Fantasy is my country. My imagination lives there."
— Susan Cooper, author of the Dark Is Rising series

"If I have something that is too difficult for adults to swallow, then I will write it in a book for children."
— Madeleine L'Engle, author of *A Wrinkle in Time*

"Reading is the best way, maybe the only way, to learn to write well."
— Lois Lowry, author of *The Giver*

"Most of the basic material a writer works with is acquired before the age of fifteen."
— Willa Cather, author of *My Ántonia*

"If there's a book you really want to read, but it hasn't been written yet, then you must write it."
— Toni Morrison, author of *The Bluest Eye*

E. Nesbit
Known for: *Five Children and It*
Full name: Edith Nesbit

P. L. Travers
Known for: *Mary Poppins*
Full name: Pamela Lyndon Travers

FIVE GIRL WRITERS TO KNOW

Phillis Wheatley (1753?–1784) was sold as a slave to a
Boston family when she was six years old. She was taught
to read, and by the time she was thirteen she had written
her first poem. Although she spoke no English when she
was brought to the New World, she quickly became fluent,
and by the age of fifteen she was reading Latin as well.
Wheatley was the first black woman to publish poetry in
America. Her book, *Poems on Various Subjects, Religious and
Moral,* was published in 1773.

Louisa May Alcott (1832–1888) was so determined to
help her family financially that she taught school, sewed,
and hired herself out as a houseworker when she was a
teenager. But her passion was writing, and at eighteen she
sold her first poem, "Sunlight." In the same year (1851),
she began to sell short stories for five dollars apiece, using
the money to support her family. She is most famous for
her novel *Little Women.*

American poet **Gwendolyn Brooks** (1917–2000) began
writing at a very young age. Her first poem, "Eventide,"
was published when she was just thirteen; four years later
she had published almost one hundred of her poems, most
in a weekly column in the *Chicago Defender* newspaper. In

1950, she became the first African American to win a Pulitzer Prize. She won for her book *Annie Allen,* which includes a poem describing the experiences of a black girl growing up in America.

Anne Frank (1929–1945) started writing in her diary on her thirteenth birthday while she and her family were in hiding from the Nazis in Amsterdam, The Netherlands. A year later the family's hiding place was betrayed, and they were sent to concentration camps. Anne died of typhus while imprisoned. The only family member to survive was her father, who later returned to their hideout and found Anne's diary. He had it published in 1947. To date, more than thirteen million copies have been printed in more than fifty languages.

Like Anne Frank, **Zlata Filipovic** (b. 1982) of Sarajevo kept a wartime diary. Zlata started her diary when she was eleven, writing at first about schoolwork and birthday parties. When Serbian troops invaded her country, her diary entries changed to stories of bombings and death. Zlata's diary was bought by a French publishing company that helped her family escape from Sarajevo. In 1994, *Zlata's Diary* was published in the United States.

PRIZEWINNERS

There have been many successful female writers throughout history. Here are some who won prizes.

Female Nobel Prize Winners for Literature

1909 Selma Lagerlöf of Sweden
1926 Grazia Deledda of Italy

1928 Sigrid Undset of Norway
1938 Pearl S. Buck of the United States
1945 Gabriela Mistral of Chile
1966 Nelly Sachs of Sweden
1991 Nadine Gordimer of South Africa
1993 Toni Morrison of the United States
1996 Wislawa Szymborska of Poland

Female Pulitzer Prize Winners for Poetry

1918 Sara Teasdale for *Love Songs*
1919 Margaret Widdemer for *Old Road to Paradise*
1923 Edna St. Vincent Millay for *The Ballad of the Harp-Weaver; A Few Figs from Thistles; Eight Sonnets in American Poetry, 1922, A Miscellany*
1926 Amy Lowell for *What's O'Clock*
1927 Leonora Speyer for *Fiddler's Farewell*
1935 Audrey Wurdemann for *Bright Ambush*
1938 Marya Zaturenska for *Cold Morning Sky*
1950 Gwendolyn Brooks for *Annie Allen*
1952 Marianne Moore for *Collected Poems*
1956 Elizabeth Bishop for *Poems — North & South*
1961 Phyllis McGinley for *Times Three: Selected Verse from Three Decades*
1967 Anne Sexton for *Live or Die*
1973 Maxine Winokur Kumin for *Up Country*
1984 Mary Oliver for *American Primitive*
1985 Carolyn Kizer for *Yin*
1987 Rita Dove for *Thomas and Beulah*
1991 Mona Van Duyn for *Near Changes*
1993 Louise Gluck for *The Wild Iris*
1996 Jorie Graham for *The Dream of the Unified Field*
1997 Lisel Mueller for *Alive Together: New Selected Poems*

Female Pulitzer Prize Winners for Fiction

1921 Edith Wharton for *The Age of Innocence*
1923 Willa Cather for *One of Ours*

1924 Margaret Wilson for *The Able McLaughlins*

1925 Edna Ferber for *So Big*

1929 Julia Peterkin for *Scarlet Sister*

1931 Margaret Ayer Barnes for *Years of Grace*

1932 Pearl S. Buck for *The Good Earth*

1934 Caroline Miller for *Lamb in His Bosom*

1935 Josephine Winslow Johnson for *Now in November*

1937 Margaret Mitchell for *Gone With the Wind*

1939 Marjorie Kinnan Rawlings for *The Yearling*

1942 Ellen Glasgow for *In This Our Life*

1961 Harper Lee for *To Kill a Mockingbird*

1965 Shirley Ann Grau for *The Keepers of the House*

1966 Katherine Anne Porter for *The Collected Stories of Katherine Anne Porter*

1970 Jean Stafford for *Collected Stories*

1973 Eudora Welty for *The Optimist's Daughter*

1983 Alice Walker for *The Color Purple*

1985 Alison Lurie for *Foreign Affairs*

1988 Toni Morrison for *Beloved*

1989 Anne Tyler for *Breathing Lessons*

1992 Jane Smiley for *A Thousand Acres*

1994 E. Annie Proulx for *The Shipping News*

1995 Carol Shields for *The Stone Diaries*

2000 Jhumpa Lahiri for *Interpreter of Maladies*

Mythology

Some myths that you know today may have been around for hundreds, or even thousands, of years. Although myths are often entertaining, they did not originate just for entertainment. Unlike folklore or fables, myths were once believed to be true. Myths helped to explain human nature and the mysteries of the world to ancient societies. Women were often key figures in these explanations.

THE OLYMPIAN GODDESSES

According to Greek mythology, twelve gods and goddesses ruled the universe from atop Greece's Mount Olympus. These Olympians had come to power after their leader, Zeus, overthrew his father, Kronos, leader of the Titans. All the Olympians were related to one another. The Romans adopted most of these Greek gods and goddesses but gave them new names.

Hera (Roman name: Juno) was goddess of marriage and the queen of Olympus. She was the wife and sister of Zeus, the sky god; many myths tell of how she sought revenge when Zeus betrayed her with his lovers. Her symbols include the peacock and the cow.

Aphrodite (Roman name: Venus) was the goddess of love and beauty and the protector of sailors. Some myths say she was the daughter of Zeus and the Titan Dione, and others say she rose from the sea on a shell. Her symbols include the myrtle tree and the dove.

Artemis (Roman name: Diana) was the goddess of the hunt and the protector of women in childbirth. She hunted with silver arrows and loved all wild animals. Artemis was the daughter of Zeus and Leto, and the twin of Apollo. Her symbols include the cypress tree and the deer.

Athena (Roman name: Minerva) was the goddess of wisdom. She was also skilled in the art of war and helped heroes such as Odysseus and Hercules. Athena sprang full grown from the forehead of Zeus and became his favorite child. Her symbols include the owl and the olive tree.

Hestia (Roman name: Vesta) was the goddess of the hearth (a fireplace at the center of the home). She was the most gentle of the gods, and she does not play a role in many myths. Hestia was the sister of Zeus and the oldest of the Olympians. Fire is one of her symbols.

Demeter (Roman name: Ceres) was the goddess of the harvest. The word "cereal" comes from her Roman name. She was the sister of Zeus. Her daughter, Persephone, was forced to live with Hades each winter; during this time Demeter let no crops grow. One of her symbols is wheat.

FEMALE MONSTERS
IN GREEK MYTH

The **Gorgons** were three horrifyingly ugly sister monsters who lived at the edge of the world. Their names were **Stheno, Euryale,** and **Medusa.** Their hair was made of serpents, and one look from a Gorgon's eyes would turn a man to stone. The hero Perseus killed Medusa by beheading her while looking only at her reflection.

The powerful monsters **Scylla** and **Charybdis** lived opposite each other in a narrow strait of ocean. Scylla had many fierce dog heads and ate sailors alive; Charybdis created whirlpools by sucking in and spitting out seawater. Their torment of the hero Odysseus is related in the *Odyssey.*

The **Sirens** were giant, winged creatures with the heads of women. They lived on rocks on the sea, where their beautiful singing lured sailors to shipwreck. Odysseus filled his sailors' ears with wax so that they might sail safely past the Sirens.

THE NINE MUSES

The nine Muses were Greek goddesses who ruled over the arts and sciences and offered inspiration in those subjects. They were the daughters of Zeus, lord of all gods, and Mnemosyne, who represented memory. Memory was important for the Muses because in ancient times, when there were no books, poets had to carry their work in their memories.

Calliope was the muse of epic poetry.

Clio was the muse of history.

Erato was the muse of love poetry.

Euterpe was the muse of music.

Melpomene was the muse of tragedy.

Polyhymnia was the muse of sacred poetry.

Terpsichore was the muse of dance.

Thalia was the muse of comedy.

Urania was the muse of astronomy.

OTHER GODDESSES
AROUND THE WORLD

North America

Many Indian tribes believe that life originated from females. Many also believe that all spirits that are life-giving forces, such as rain and corn, come from female deities.

Sedna rules over the sea animals. The Inuits (Eskimos) believe that she uses ugliness as protection. Anyone who dares to look at her will be struck dead.

Selu, the corn mother of the Cherokee, cuts open her breast so that corn can spring forth and give life to the people.

Blue Corn Woman and **White Corn Maiden** are the first mothers of the Tewa Pueblo people. Blue is the summer mother; White is the winter mother.

The Three Sisters, in the Iroquois tradition, give the life-supporting forces of corn, beans, and squash. The Three Sisters are thanked daily.

White-Painted Mother is the mother of Child of the Water, from whom all Apaches are descended. She keeps her child safe in her womb, slays all evil monsters, and keeps the world safe for Apaches.

White Buffalo Calf Woman, for the Lakota Indians, is the giver of the pipe. The pipe represents truth.

Ancient Mexico

Chalchiuhtlicue was the Aztec goddess of all waters on earth, but she was especially associated with running water. She was frequently depicted as a river that nourished a prickly pear tree.

Chicomecoatl was the goddess of corn and fertility. So

important was corn to the Aztecs that she was also known as the goddess of nourishment. She was sometimes depicted carrying the sun as a shield.

Coatlicue was the goddess of the earth and mother of all the gods. She also gave birth to the moon and the stars. She was depicted wearing a skirt made of snakes.

Xochiquetzal was the goddess of flowers, dance, and love. Birds and butterflies loved her and were frequently in her company.

China

Ma-Ku personifies the goodness in all people. She took land from the sea and planted it with mulberry trees. She freed the slaves from her cruel father.

Kuan Yin represents wisdom and purity for the Chinese. She has a thousand arms, symbolizing her infinite compassion.

Ancient Egypt

Isis invented agriculture. She was the goddess of law, healing, motherhood, and fertility. She was sometimes depicted with a disk from the sun between two cow horns on her head, or with a headdress in the shape of a vulture.

Hathor, the goddess of love and mirth, protected children and pregnant women. She embodied the sky and was often depicted as a star-speckled "celestial cow," or just with a cow's head or cow horns.

Nephthys was the goddess of the dead. She was a kind and understanding companion to the newly dead as well as to those left behind. She was sometimes represented as a bird, or with hieroglyphics that spelled her name above her head.

Nut represented the heavens and helped to put the world in order. She had the ability to swallow stars and the pharaohs and cause them to be born again. Her body was covered with painted stars. She existed before all else had been created.

Hawaii

Pele is the powerful Hawaiian goddess of fire. She lives in the Kilauea Volcano and rules over the family of fire gods. When she is angry she erupts and pours fiery rock over the land.

Hiiaka is the youngest sister of Pele. She is a fierce warrior and yet a kind and calm friend of humanity. She gave people the healing arts, the creative arts, and the gift of storytelling.

Ancient Ireland

Danu was the goddess of the earth. She was mother of many important gods, as well as mother of the Tuatha De Danann, a mythological race of people who had great and magical powers.

Brigid was a threefold goddess. She ruled poetry and inspiration; healing and medicine; and war and weapons crafting.

Cerrid was the Irish goddess who gave intelligence and knowledge to humans.

Caillech was the wisest woman. She ruled the seasons and the weather and was believed to have the power to move mountains.

Ancient Yucatan

Ixchel was the Mayan goddess of the moon and the protector of pregnant women. She was often depicted as an old woman wearing a full skirt and holding a serpent.

Ixtab was the goddess of the hanged. The Mayans believed that those who died by hanging went to paradise. Ixtab was depicted with a noose around her neck.

Ancient Scandinavia

Freyja was the goddess of love and fertility. She was very beautiful and enjoyed music and song. Fairies were among her most beloved companions.

Frigg was the goddess of the sky, marriage, and motherhood. It was believed that she knew the fate of each person but kept it a close secret.

Hel was the goddess of the dead and queen of the underworld. She was hideous and bad tempered; her body was in partial decay.

The Norns were three sisters who lived around the tree of life. They controlled fate: Urth ruled the past, Verthandi ruled the present, and Skuld ruled the future.

Ancient Sumer

Ereshkigal was the goddess of the underworld. She was stubborn and temperamental and could be difficult to please. However, hymns and offerings to the dead could help to improve her mood.

Inanna was the goddess of love, fertility, and war. She was sometimes called "Queen of the Sky," and her symbol was an eight-pointed star. Inanna was also queen of all beasts and counted the lion as her special companion.

Ninhursag was the goddess of the earth and mother of the gods, as well as the creator of all plant life. When she was pleased, the ground was fertile and the seasons were rich; but when she was angry, fields were barren.

MORE ABOUT MYTHS

Would you like to learn more about mythology? Try some of these books.

American Indian Myths and Legends, edited by
Richard Erdoes and Alfonso Ortiz
Suitable for all ages, this abundant work brings together more than a hundred tales from dozens of American Indian traditions.

D'Aulaire's Greek Myths, by Ingri and Edgar Parin D'Aulaire
This classic work — perhaps the all-time finest book of mythology for children — pairs tales of love, magic, greed, and trickery with unforgettable illustrations.

D'Aulaire's Norse Gods and Giants, by Ingri and
Edgar Parin D'Aulaire
Here the D'Aulaires bring their gifts for storytelling and art to vivid, thrilling, and sometimes funny renderings of Scandinavian myth.

Gods and Pharaohs from Egyptian Mythology,
by Geraldine Harris
A collection of related Egyptian myths, this book features brilliant illustrations and a brief introduction to hieroglyphics.

The Names upon the Harp: Irish Myth and Legend,
by Marie Heaney
This book collects nine tales from early Celtic mythology and related legends. Be warned: the gripping exploits of these heroes and heroines may keep you up at night.

Friendship

LANDMARKS

1874 The Philadelphia chapter of the YWCA hosts its first summer camp, a "vacation project" for girls who are tired out from working long hours.

1902 Schoolteacher Josephine "Jessie" Field Shambaugh, "the Mother of 4-H," starts the Girls Home Club in Page County, Iowa, to help girls learn practical skills. Ten years later, combined with her other venture, the Boys Corn Club, it evolves into 4-H.

1910 The first Camp Fire Girls meetings, organized by Luther and Charlotte Gulick, are held in Vermont.

1912 The first Girl Scout troop in America is founded in Savannah, Georgia, by Juliette Gordon Low.

1928 The World Association of Girl Guides and Girl Scouts is formed to bring together girls from more than one hundred countries.

1936 The Student Letter Exchange, based in Lynbrook, New York, is organized to start pen pal relationships among students in different countries.

1979 The youth organization Big Brothers of America introduces its girls' program, Big Sisters of America.

1997 The introduction of AIM (AOL Instant Messenger) makes "live chat" possible for friends at different computers.

FUN WITH YOUR FRIENDS

True friends are loyal and trusting and have shared interests. True friends have fun together. Here are some fun
things you and your friends can do.

Make up a treasure hunt. You can divide into two teams.
One team hides secret clues — using pictures or words —
or makes a treasure map, and the other team hunts for the
treasure.

Publish a newsletter about your lives. Include news stories about what you have been up to lately — hobbies or
school events — and drawings or photographs. Send
copies to friends and family who don't live close by.

Cook a theme meal for friends and family. You could do
a pizza theme, with each person using toppings to decorate a pizza. You could do a small food theme, serving
things like small fruits and sandwiches cut into tiny pieces
on tiny plates.

Put on a play. A book you have all read can provide the
story for a show. Besides performing, you can work
together to find props, put together a set, and think of
costumes. You can also make a program guide that tells
about your play.

Throw a nature party. In spring or summer, you can plant
seeds. In fall, you can collect leaves or pinecones. In winter,
you can make snow angels and snow sculptures. Birdseed
makes a good party favor for wildlife at any time of year.
Write a never-ending story. In a notebook, write one paragraph or chapter of a story. Then pass it around, with each
friend adding another paragraph or chapter. You'll be surprised at how it all turns out!

JUMP ROPE JIVE

About a hundred years ago most jump rope jumpers were boys. Each boy had his own rope and jumped in competition with other boys. When girls took to jumping rope, they made it a group game among friends and added rhythm and song. Here are some jump rope songs:

Two in the middle and two at the end,
Each is a sister and each is a friend.
A penny to save and a penny to spend,
Two in the middle and two at the end.

Candy, candy in a dish.
How many pieces do you wish?
1, 2, 3 . . .

Apple on a stick,
Five cents a lick.
Every time I turn around
It makes me sick.

Sugar, salt, pepper, cider.
How many legs has a bow-legged spider?
1, 2, 3 . . .

Rooms for rent,
Inquire within.
As I move out,
Let _____ come in.

All in together, girls.
How do you like the weather, girls?
January, February, March, April . . .
(Jumpers run in on their birthday months.)

STORY FRIENDS

If you would like to read some great stories about friendship, try these books.

Anne of Green Gables, by L. M. Montgomery, tells the story of two girls, Anne Shirley and Diana Barry, who become strong friends.

Rebecca of Sunnybrook Farm, by Kate Douglas Wiggin, tells the story of a friendship between Rebecca Randall and Emma Jane Perkins.

Jennifer, Hecate, Macbeth, William McKinley, and Me, Elizabeth, by E. L. Konigsburg, is the unusual tale of two lonely girls who forge a strong and sometimes painfully honest friendship while studying to be witches.

Charlotte's Web, by E. B. White, tells of the friendship that grows between Wilbur, a runty pig, and Charlotte, a heroic spider.

The Betsy-Tacy books, by Maude Hart Lovelace, chronicle the friendship of Betsy Ray, Tacy Kelly, and Tib Muller from when they are little girls, through their high school years, and until Betsy marries.

FIVE FRIENDSHIPS TO KNOW

Ruth and **Naomi** are figures from the Hebrew Bible. Naomi left her home in Bethlehem with her husband and sons because of a famine. They moved to Moab, where her sons married Ruth and Orpha. When Naomi's husband and sons died, Orpha stayed in Moab, but Ruth would not leave her mother-in-law, who was returning to Bethlehem. Ruth's faithfulness to Naomi has become a famous example of loving friendship.

Anne Sullivan met **Helen Keller** when she was hired to teach her. Sullivan was twenty, Keller was seven. In 1904 Keller wrote *The Story of My Life,* in which she said that Sullivan transformed her from an angry child into the best-educated blind and deaf person in the world. Their teacher-student relationship evolved into a true friendship: they lived together for forty-nine years, until Sullivan died in 1936.

Gertrude Stein's first commercial success was her book *The Autobiography of Alice B. Toklas.* Toklas was Stein's best friend for nearly forty years. In the book Stein describes how Toklas was her proofreader, screener of visitors (of which she had many), and confidante. Toklas was also Stein's nurse when she became ill and died in 1946.

When feminists **Elizabeth Cady Stanton** and **Susan B. Anthony** met in 1851, they became close friends. Working for the same cause, women's rights, they thought alike,

---

🖎 SHE SAID IT 🖎

"Walking with a friend in the dark is better than walking alone in the light."
— Helen Keller, American writer and educator

"Winning has always meant much to me, but winning friends has meant the most."
— Babe Didrikson Zaharias, American golfer and all-around star athlete

"It seems to me that trying to live without friends is like milking a bear to get cream for your morning coffee. It is a whole lot of trouble, and then not worth much after you get it."
— Zora Neale Hurston, American folklorist and author

"Friendship with oneself is all-important because without it one cannot be friends with anyone else in the world."
— Eleanor Roosevelt, American first lady

"Advice is what we ask for when we already know the answer but wish we didn't."
— Erica Jong, American author

---

worked together, and became the bedrock of the women's suffrage movement.

Anne Bonny and **Mary Read** were pirates and best friends. They met when Read was captured and taken aboard Bonny's pirate ship. Both had been disguised as men, but when they discovered they were both masquerading, they became inseparable friends. They were captured and tried as pirates in 1720 in Jamaica. We don't know what happened to Bonny, but Read died of a fever in prison.

FRIENDSHIP TROUBLES

Friends can help you sort out your troubles, but sometimes you can run into troubles with your friends. Here are some difficult situations you might face.

Jealousy

If something nice happens to your friend, but you don't feel glad about it, you might be feeling jealous. Jealousy is a very common emotion that people can have at all ages. If you feel jealous, that's okay, but don't let it get in the way of your friendship. It can help to take some time away from your friend to think about why you feel bad.

Your Best Friend "Drops" You

The school year has just started, but your best friend isn't acting like a best friend at all. She's spending all her time with someone she met over the summer. What can you do? Friendship is a living, growing thing, and sometimes our friends find other people with whom they want to spend time. Sometimes, all of a sudden, a friend will meet someone else whom she just really likes.

As bad as it feels, there's probably nothing you can do. Your best friend might want to be friends again later, or she might not. You might not be friends again in the same way. No matter what happens, it is important to remember that you are still a good person and a good friend. And you will find other best friends.

Your Friend's Loved One Dies

It can be hard to know what to say or how to act around a friend whose loved one has died. Many adults also feel unsure about what to do. Remember that your friend is still your friend, even if she cries a lot or is quiet or angry or just seems different. Continue being her friend. If you don't know what to say, that's fine. You can just say, "I'm sorry." You can ask, "How are you feeling?" You can say, "I care about you." Or you can just be quiet and listen. Your friend may want to talk about the person who has died.

When someone we love dies, we can feel bad for a very long time. One of the most important things a friend can do is stick with someone who's grieving even when that person doesn't seem like herself. The death of a loved one is one of the hardest things that can happen to a person at any age. Having good friends can make it a little bit easier to bear.

Fashion

LANDMARKS

1470 To hide her pregnancy, Queen Juana of Portugal wears the first hoop skirt.

1477 Anne of Burgundy becomes the first woman to receive a diamond engagement ring. It is given to her by Maximilian I of Germany.

1850s In her magazine, *The Lily,* American feminist Amelia Bloomer promotes the comfort of "bloomers," Turkish-style trousers worn under a simple flaring skirt.

1873 Jeans are first made by Levi Straus, an American.

1913 French designer Gabrielle "Coco" Chanel opens her boutique in Deauville, France. Her chic and comfortable knit suits herald the modern era of women's fashion.

1922 The first "flesh-colored" stockings are sold for white women. It will be another twenty-five years before stockings for women of color are available.

1926 Hemlines rise to knee height for the first time.

1938 Nylon stockings are invented. They are first sold in stores in 1940.

1940 Shoulder bags for women first appear as part of service uniforms worn during World War II.

1955 London designer Mary Quant opens Bazaar, her Carnaby Street boutique. Her miniskirts, tights, and crocheted tops will define the new "youth culture" look.

1960s Fun furs of acrylic and polyester are made to look like real fur.

1978 The first designer jeans are fashioned by Gloria Vanderbilt.

1981 The Malden Mills company invents Polarfleece. The soft, quick-drying fabric, made partly from recycled plastic, makes bundling up cozier than ever.

1984 Through music videos, Cyndi Lauper shows the world her wacky, colorful look — part vintage, part punk — and helps make thrift shopping the new frontier of chic.

1990s Counterfeit clothing is made by computer, producing brand-name fakes. Polo, Guess?, Gap, Banana Republic, DKNY, and Disney are all copied.

A CENTURY OF U.S. FASHION

1900s

upswept hair	feathered hats
tight collars	lots of lace
corsets for tiny waists	skirts with trains
lace-up boots	

Icon: The Gibson Girl, created by illustrator Charles Dana Gibson. This elegant young lady wore a floor-length skirt and a starched, tailored shirt with an ascot tie at the neck.

1910s

decorated hairpins	beaded handbags
draping blouses	sashes

narrow "hobble" skirts   Middle Eastern patterns
bows on shoes

Icon: The modern girl, who took off her corset and revealed a more natural shape. World War I fabric shortages made her practical, in a simply cut dress of plain, dark material.

1920s

bobbed hair   cloches (close-fitting hats)
long necklaces   "flapper" dresses
sheer stockings   T-strap shoes
coats that fastened on the side

Icon: The American flapper. She had short hair and wore a sleeveless, drop-waist dress. To complete the outfit she wore ropes of pearls around her neck and a cloche.

1930s

hats worn at an angle   fur collars
shoulder pads   patterned sweaters
long, flowing gowns   sandals
one-piece wool bathing suits

Icon: Nancy Drew. The stylish sleuth first appeared in books and movies in the 1930s, wearing gently tailored suits, ruffled blouses, dainty heels, and jaunty hats.

1940s

the pageboy haircut   brooches
halter tops   cardigan sweaters
rolled-up blue jeans   sleek evening dresses
cork-soled "wedgie" shoes

Icon: Rosie the Riveter. A symbol of the many women who took factory jobs during World War II, she sported overalls and a "'do-rag" kerchief that protected her hairstyle.

1950s

ponytails	black leotards
poodle skirts	petticoats
strapless evening gowns	ballerina flats
pedal pushers (calf-length pants)	

Icon: The bobby-soxer, whose look is still popular for Halloween costumes. A circle skirt, a neck scarf, and bobby socks with saddle shoes are the trademarks of this style.

1960s

the beehive hairdo	white vinyl "go-go" boots
peace signs	paisley and Indian prints
bell bottoms	miniskirts
pale lipstick and dark eyeliner	

Icon: The hippie or flower child, a look that started out as fashion rebellion. The costume often included tie-dyed shirts, old jeans, "love beads," and peace symbols.

1970s

the Afro	lots of lip gloss
denim jackets	pantsuits
earth tones	wraparound skirts
western boots	

Icon: Annie Hall, as played by Diane Keaton in the movie *Annie Hall.* She looked casual and uncoordinated in baggy pants, floppy hats, and loose-fitting men's shirts with ties.

1980s

big hair with lots of mousse	
bangle bracelets	Fair Isle sweaters
acid-wash denim	leg warmers
ankle socks	penny loafers

Icon: The preppy, who wore polo shirts, pearl chokers, khaki pants, and loafers. This conservative look imitated the kind of clothing worn by preparatory school students.

1990s

colored hair
puffy jackets
bare midriffs
designer sneakers

mehndi (henna tattoos)
hooded sweatshirts
tracksuit pants

Icon: The homegirl, whose urban hip-hop style became a global trend. In a baby tee, baggy jeans, and name-brand sneakers, the homegirl was all that and a bag of chips.

COLOR AND FASHION

People have always used color to suggest a mood or a meaning. Here are some of the ways that fashion and color have connected.

The Aztecs of Mexico taught the Spanish how to make **red** dye by crushing insects called cochineals. Deep red looks bold, while pale red — pink — looks gentle. In the United States, pink is now associated with girls, though before the 1920s it was considered a boys' color.

Cheerful, sunny **yellow** gets noticed. In ancient Rome, yellow was the most popular wedding color. Yellow is sometimes worn for safety reasons: raincoats today may be bright yellow so that the wearer can be seen easily in the rain.

Blue is the most common clothing color — especially since blue jeans are everywhere! Blue has a calming effect. Fashion consultants recommend wearing blue to job interviews because it symbolizes loyalty. For this same reason, U.S. police officers traditionally wear blue.

Brilliant, tropical **orange** may be the most attention-grabbing color. Like yellow, it can be used for safety: evening joggers and hunters in the woods may wear orange tops. Orange can be mixed with pink to make peach, a lively color that is popular in warm weather.

Green is the easiest color on the eye. Hospital uniforms may be green because the color relaxes patients. Green is also associated with nature; leprechauns are said to dress in green. Brides in Europe in the Middle Ages wore green to symbolize fertility.

Purple has always been considered the color of royalty, because for a long time it was very rare. Cleopatra needed twenty thousand snails soaked for ten days to obtain one ounce of purple dye for her royal clothing.

Black is generally considered a serious color. In the West, black is traditional for both funeral dress and sophisticated evening wear. Black outfits can be overpowering, and villains, such as Dracula, often wear black.

Since the twentieth century, Western brides have worn **white** to symbolize purity. In China, however, white is the color of mourning. White shows dirt easily; doctors and nurses wear white coats to show that they understand that cleanliness is important.

FOOTWEAR FACTS

- Sandals originated in warm climates where the soles of the feet needed protection but the tops of the feet needed to be cool.

- Four thousand years ago the first shoes were made of a single piece of rawhide that enveloped the foot for both warmth and protection.

- In Europe, pointed toes on shoes were fashionable from the eleventh to the fifteenth centuries.

- In the Middle East, heels were added to shoes to lift the foot off the hot sand.

- In Europe in the sixteenth and seventeenth centuries, heels on shoes were always colored red.

- All over the world, pairs of shoes were identical until the nineteenth century, when left- and right-footed shoes were first made in Philadelphia.

- In Europe, it wasn't until the eighteenth century that women's shoes were different from men's.

- Six-inch-high heels were worn by the upper classes in seventeenth-century Europe. Two servants, one on either side, were needed to hold up the person wearing the high heels.

- Boots were first worn in cold, mountainous regions and in hot, sandy deserts where horse-riding communities lived. Heels on boots kept feet secure in the stirrups.

- The first lady's boot was designed for Queen Victoria in 1840.

- Wearing high heels has made it difficult for women to move quickly and has resulted in bunions, corns, twisted ankles, spinal deformities, and shortened calf muscles.

UNDER IT ALL

Corsets

Corsets and girdles were first worn outside of clothing. This practice is evident in many European national costumes, such as that of Bavaria.

DID YOU KNOW?

Haute Couture — at Haute Prices

The term "haute couture" (say "OAT coo-TOAR") is French. *Haute* means "high" or "elegant." *Couture* literally means "sewing," but the term has come to indicate the business of designing and creating custom-made, high-fashion women's clothes. To be called an haute couture house, a business must belong to the Syndical Chamber for Haute Couture in Paris. The syndicate has about eighteen members, including such fashion giants as Coco Chanel, Christian Dior, and Pierre Cardin.

Made from scratch for each customer, haute couture clothing typically requires three fittings. It usually takes from one hundred to four hundred hours to make a single dress, which costs from $26,000 to more than $100,000. A tailored suit starts at $16,000, an evening gown at $60,000.

The ancient Greeks were the first to wear girdles. They called them "zones." A band of linen or soft leather was bound around a woman's waist and lower torso to shape and control her body.

In Florence during the Renaissance, the noblewoman Catherine de Médicis decreed it bad manners to have a thick waist. She helped to design a hinged corset that narrowed the waist to thirteen inches.

The iron corset was devised in 1579 and was worn by women for about ten years.

The first modern corset was made in Britain in the 1700s; its tight lacing made breathing and movement difficult.

A shorter, lighter corset was made in America in 1911 so women could have the freedom of movement to dance the tango.

Underpants

European women did not wear underpants until the early 1900s.

As the twentieth century began, most Americans wore union suits or "all-in-ones" — undergarments that combined pants and a top.

In the 1930s, Americans traded their union suits for separates, and the word "underpants" entered dictionaries. The 1930s also saw another major change in underwear: easy elastic waists replaced button, snap, and tie closures.

"Day of the week" underpants were a craze in the 1950s. Each pair of underpants in the set of seven was labeled with a different day of the week.

Colorful Underoos hit stores in 1978. The fun underwear secretly transformed thousands of girls into Wonder Woman.

Dressing Your Breast Friends

Statues found on the Greek island of Crete dating from 2500 B.C. show women wearing bralike corsets that lifted their breasts out of their clothing.

American Marie Tucek patented the first bra, which she called a "breast supporter," in 1893. It had two supportive cups and shoulder straps.

In 1913, New Yorker Marie Phelps Jacob fashioned a flattening bra from two handkerchiefs and some ribbon. The following year she patented her invention.

In the 1920s, women wore bras that pressed down their breasts for the then-popular flat-chested look.

Also during the 1920s, a Russian immigrant named Ida

Rosenthal founded the Maidenform lingerie company with her husband, William. They made bras for women of every size and introduced the cup system (A, B, C, D).

Pointy, cone-shaped bras were a trend in the 1950s.

In recent years, some bras have included little pillows of gel or water to make the chest look larger. Others can be pumped up with a little air pump.

Today, the average American woman owns six bras. White is the best-selling color for bras.

THE WORLD OF FASHION

The next time you put on your cologne or bikini, remember that they took their names from real places on the globe.

Bikini, a tiny coral island in the Pacific's Marshall Islands, is where the United States conducted atom bomb tests in the late 1940s. Four days after the A-bomb was detonated, a French designer introduced a scanty, two-piece bathing suit and called it the "bikini." He believed it would cause a fashion explosion, and indeed it did.

Cologne (spelled **Köln** in German) is the city in Germany where cologne was first produced. Cologne is a scented liquid made of alcohol and various fragrant oils, similar to perfume.

Nîmes, France, is the source of denim. In French the material was called *serge de Nîmes,* or "fabric from Nîmes," and "de Nîmes" became "denim."

Suede comes from the French pronunciation of "Sweden," the country where this soft, velvety leather was first made.

FIVE WOMEN TO KNOW

Jeanne Lanvin (1867–1946) was famous for creating the unfitted, chemise-style flapper dress. She first designed the loose, flowing dress for her little daughter, and when its freshness and simplicity were admired by women of all ages, she adapted the design for adults. Lanvin was also known for using beautiful, subtle pinks and blues that were inspired by her own magnificent collection of oil paintings. Her House of Lanvin, the oldest of all Paris fashion houses, set the standard for fashion in the 1920s.

Gabrielle "Coco" Chanel (1883–1971) is considered the most significant designer of the twentieth century. She introduced the little black dress, sweater sets, the pleated skirt, triangular scarves, and fake pearl necklaces; she pioneered the use of knit jersey as a fashion fabric; and she produced the first artificial suntan lotion. She manufactured her own perfume, which she called Chanel No. 5, after her lucky number. Chanel was also famous for changing black from a color of mourning to a color of elegance. It is said that Coco Chanel knew what women wanted to wear before they knew it themselves.

Elsa Schiaparelli (1890–1973) became famous for designing innovative, sophisticated clothes. One of her most memorable designs was a tall hat in the shape of a shoe. Her dresses and suits of the 1930s with squared, padded shoulders changed the female figure. She originated the idea of separates for sports clothes. She was the first designer to use zippers and man-made fabrics for her fashions. Schiaparelli was also known for her bold use of color, particularly shocking pink and ice blue, both of which became sensations after she introduced them.

Mary Quant (b. 1934) turned the fashion world upside down with her outrageous styles. She is credited with creating both the mod look of the 1960s, with its futuristic vinyl clothing and bold geometric designs, and the miniskirt. To complete the look with smoky eyes and pale lips, Quant introduced her own line of cosmetics. In 1966, Quant was awarded the Order of the British Empire, a medal of achievement given by the queen, for her contribution to fashion. She accepted this honor wearing a miniskirt.

Diane Von Furstenberg (b. 1946) is the creator of the now classic V-neck wraparound knit dress. During the 1970s, when Von Furstenberg introduced this slinky dress, she was selling about twenty thousand a week. By 1975, more than five million had been sold. Her wraparound dress had a revival in the late 1990s, when Von Furstenberg brought it back by popular demand. Once again, the casual, easy-to-wear dress was a hit, and it quickly sold out in many stores.

FAMOUSLY UNFASHIONABLE

There have always been some women who couldn't care less about fashion.

British author **Iris Murdoch** wore ratty canvas tennis shoes almost all the time, even when she went to Buckingham Palace to receive an award from the queen. Around the campus of Oxford University, where she taught for many years, Murdoch was known as a remarkable slob, with baggy old clothes and shaggy hair that looked like it had been cut under a bowl. (Nevertheless, she was also known for having had many boyfriends.)

From Head to Toe

Young girls living in the United States two hundred years ago were expected to cover themselves from head to toe — twelve layers, to be exact. And you thought it took *you* a long time to get dressed! Here's a look at what they wore.

- a frilly undershirt
- a bodice with lots of buttons
- a garter belt to hold up stockings
- long stockings
- long underpants that were buttoned to the bodice
- shoes
- a red flannel petticoat
- a starched petticoat
- a dress
- an apron
- a hair ribbon
- a bonnet

Humanitarian and former first lady **Eleanor Roosevelt** was another woman whose mind was on other things. In an era when it was essential for women to look ladylike, Roosevelt wore her riding clothes to meet with journalists. When a newspaper article included speculation about just how little the wealthy but casual first lady spent on her wardrobe, Roosevelt considered it a high compliment and saved the article in her scrapbook.

Although she was a child model, and starred as the fashion-obsessed teenager Cher in the movie *Clueless*, actress

Alicia Silverstone does not like to think about clothes. "I hate clothes, I hate fashion!" Silverstone has said, adding that she much prefers walking her dogs or working in her garden. When she has to dress up for awards ceremonies or other high-profile events, she asks a stylist to find an outfit for her so that she doesn't have to concern herself.

ALL MADE UP

Women and men have always used paints, powders, dyes, and perfumes to decorate their hair, faces, and bodies. From earliest times, people have used colorful makeup to frighten enemies, to show social rank, for religious ceremonies, in puberty rites, to make magic, to protect the skin and eyes, and to make themselves attractive to mates.

Ancient Egypt and Rome

- Women and men both used rouge, lipstick, and nail polish.

- Black and green eye shadow was used to protect the eyes from the desert sun.

- Women traced the veins in their skin with blue paint.

- Black kohl was used as mascara, eyebrow darkener, and eye liner.

- Body moisturizers included sesame, olive, palm, and almond oils.

- Perfumes were made of musk, thyme, myrrh, and frankincense.

- Hair dyes were made from henna, from the blood of black cows, and from crushed tadpoles in warm oil.

"Fashion is made to become unfashionable."
— Coco Chanel, French designer

"Fashion can be bought. Style one must possess."
— Edna Woolman Chase, American fashion editor

"A fashionable woman wears clothes; the clothes don't wear her."
— Mary Quant, British designer

"I base most of my fashion sense on what doesn't itch."
— Gilda Radner, American comedienne

"My weakness is wearing too much leopard print."
— Jackie Collins, British novelist

- The first frosted look in makeup was achieved by pulverizing ant eggs and adding them to face paints.

- The Romans used crocodile excrement for mud baths, barley flour and butter for pimples, and sheep fat and blood for nail polish.

- Roman men and women frequently dyed their hair blond. The dyes were so caustic that many people lost their hair and had to wear wigs.

More from the History of Cosmetics

- In the Middle Ages, European society women painted their faces white or were bled (actually had some of the blood drained out of their bodies) to achieve a pale complexion.

- In China and Japan, rice powder paint was used to paint

faces white. Eyebrows were plucked, and teeth were painted black or gold.

- In Europe in the Middle Ages, beauty patches worn on the face had meaning. Adhesive fabrics cut in the shapes of stars, hearts, and crosses were worn in the following manner: one to the right of the mouth meant the woman was flirtatious; one on the right cheek meant she was married; one on the left cheek meant she was engaged; and one at the corner of an eye meant she was passionate.

- In Elizabethan England, dyed red hair was the fashion. Women also slept with slices of raw beef on their faces to reduce wrinkles.

- European men stopped using perfumes and wearing cosmetics during the Victorian era.

- Lipstick was first manufactured in the United States in 1915. By the 1930s, it was an essential. Purple was the color of the 1960s, and white lipstick was popular in the 1970s. Gloss ruled the 1980s, and very dark red was a 1990s trend.

Love and Romance

LANDMARKS

Ancient Greece In mythology, Eros, a plump, winged baby, is worshipped as the god of love. The Romans call him Cupid.

A.D. 270 The Roman priest St. Valentine is martyred on February 14 for being a Christian. According to legend, he fell in love with the daughter of his jailer, and just before his death wrote her a note signed "from your Valentine."

Middle Ages Flower girls first appear in wedding ceremonies. Two young girls — usually sisters — carry wheat before the bride in the procession. Later, flowers replaced the wheat, and it became customary for the flower girls to strew petals at the bride's feet.

1840s The first commercial Valentine cards, trimmed with imported lace, are made by Esther Howland in Worcester, Massachusetts.

1860s NECCO Sweethearts Conversation Hearts are invented. These first hearts have printed paper notes tucked inside. The lengthy, old-fashioned sayings include such wistful thoughts as "Please send a lock of your hair by return mail."

1920s The custom of wearing a new white dress for a wedding ceremony becomes the norm in much of the West. Before this time, few women could afford a dress they would wear only once, and a bride would simply wear a formal gown of any color.

SUPERSTITIONS: LOVE LORE

A superstition is the belief in the magical ability of an object or an action to influence one's life. Folklore abounds with superstitions related to love and marriage; here are some of them.

Surefire Signs You'll Fall in Love Soon

- You stumble going up a flight of stairs.
- You have hairy legs.
- You dream of taking a bath.
- The lines on your palm form an *M.*

To Dream of What Your Next Sweetheart Will Look Like

- Sleep with a mirror under your pillow.
- Wear your nightgown inside out.
- Rub your headboard with lemon peel before turning off the light.
- Count nine stars each night for nine nights.
- Put daisies under your pillow at night.
- Take a sprig of rosemary and a sprig of thyme. Sprinkle them three times with water and place each herb in a shoe. Put the shoes at the foot of your bed.
- Stand in front of a mirror and brush your hair three times.

CRUSHED OUT

People get crushes throughout their lives. Crushes can be fun. Daydreaming about your crush can feel safe and nice. But crushes can also feel, well, crushing.

If having a crush is dragging you down, but you can't seem to stop it, try this: picture your crush doing ordinary things. Envision him being obsessed with a video game, whining at his mom, or falling asleep drooling on the couch — still fascinating? In your imagination, your crush is probably perfect in almost every way. But reality can rarely live up to your imagination. That's why it's just as well that most crushes stay one-sided!

Crushes can be useful, because they can help you figure out what qualities are important to you. Does a good sense of humor get your attention? Do you like the thoughtful way your lab partner listens to you? Perhaps your friend's love of music is something that will be important to you throughout your life. You can think of a crush as practice before real love comes along.

SWEETS FOR SWEETHEARTS

For nearly a century and a half, the makers of NECCO Sweethearts Conversation Hearts have come up with some of the sweetest ways of saying "I love you." Every Valentine's Day the company presents new messages on the tiny colored hearts that have been a holiday tradition since the Civil War.

However, some favorites among the more than one hundred Sweetheart sayings have been in circulation since the hearts were first factory made in 1902. These classics include "Kiss Me," "Sweet Talk," and "Be Mine." Sometimes a motto is discontinued for a time and then re-appears; others are gone for good. Sayings considered outdated by NECCO include the funky "Dig Me" and the cheerful "You Are Gay."

If you miss some of the old sayings, or would like to see some of your own invention, you can have them custom made. The catch is that you'll have to buy a full production run, or about 1.7 million candy hearts. But you'll have plenty of time to eat them — they should stay fresh for at least five years.

MARRIAGE THROUGH TIME

It is believed that the first "marriage" took place when a primitive man went into a primitive woman's cave and carried her off to be his mate. He chose her not for love but for her ability to work and reproduce. Since then, of course, the idea of marriage has changed quite a lot.

Ancient Greeks and Spartans

All marriages were arranged by parents and approved by the gods in ancient Greece. Women in their early teens were married to men in their mid-thirties. A husband then had to buy his new wife from her father. Many couples did not see each other for the first time until after the ceremony, when the bridal veil was removed. On the night before the wedding, the girl's hair was cut off, and she was bathed in holy water from a sacred fountain. Her childhood toys were then taken away and dedicated to a goddess. Greek wives were "owned" by their husbands, who could lend or sell them to others.

The Spartans believed that a person's athletic ability matched his or her fitness for marriage. Before marrying, a couple was required to wrestle in public to show their compatibility. Spartan women married in their twenties. The groom's father chose a bride for his son. Twelve months after the selection, the couple was married. During the marriage ceremony, the bride wore a white robe, a veil, and jewelry given to her by her new husband's family.

Dowries: The Joining of Money and Marriage

A dowry — the money or property a bride brings to her husband upon marriage — was common throughout much of the ancient world and is still part of some traditional cultures today. A dowry can have various functions. A dowry of household goods can help the newlyweds set up their own home. A dowry of property or jewelry can help the wife support herself if her husband dies.

Sometimes the groom's family pays for the bride, often to compensate her family for the money spent raising her. Or, if the bride has been a valuable worker, her family may be compensated for the loss of her economic support.

The ceremony took place in the groom's tent, and the festivities lasted seven days. If a woman was wealthy, she might have a husband for each house she maintained.

Ancient Romans

Roman brides wore white tunics with orange veils and orange slippers. Following the ceremony, the groom carried his bride over the threshold of their new home to symbolize his ownership of her.

Medieval Christians

Christian church marriages were thought to be made in heaven and therefore could never be broken. The father of the bride gave a dowry of land or money to the groom. If the marriage was unsuccessful, the wife and the dowry were returned to the father's home, but neither partner was allowed to remarry.

Ancient Japanese

Until the 1400s, married couples did not live together in Japan. They stayed in separate homes, meeting only at night. The old Japanese word for marriage meant "slip into the house by night."

MARRIAGE TODAY

Amish

When an Amish couple wants to marry, the man asks a churchman to ask the woman's parents for their approval. If they consent, the marriage is announced two weeks before the wedding. The wedding takes place on a Tuesday or Wednesday in November, after the harvest. The bride wears white for the first and only time in her life. There are no rings, photographs, or flowers at the wedding. There is no honeymoon, and the couple does not live together until the springtime, after a series of weekend visits with family and friends.

Arabs

Arab marriages are arranged between two families. The families agree on the amount of money to be paid to the bride's family for her trousseau (a wardrobe the bride acquires before marriage). An Arab bride celebrates her wedding in an ancient ceremony that excludes men. The bride's hair is covered with henna, a deep red dye, and her body is elaborately painted by her friends. After the ceremony, the women all dance together.

French

In France, one couple may have three marriage ceremonies. The first is the civil ceremony, which is performed in the town hall with the mayor or a representative officiating. The second ceremony is religious, usually Roman Catholic, performed by a priest. The third takes place if the couple lives in the countryside. In this ceremony, the people of the village host a ten-course banquet for the bride and groom that includes singing, storytelling, games, and toasting. The villagers bang pots and pans to remind the couple of the possible difficulties of marriage.

Germans

In a wedding ceremony in Germany, the bride and groom hold candles decorated with ribbons and flowers.

Greeks

At a Greek wedding ceremony, a guest of honor, known as the *koumbaros,* crowns the wedding couple and joins them in a symbolic gesture by circling the altar three times.

Indians

Child marriages are still common in parts of rural India, where it is not unusual for seven-year-olds to marry! On the day of the ceremony, the young groom rides into town on a horse followed by hundreds of friends and relatives. A local wise man chants wedding mantras, or prayers. The bride and groom walk around a ceremonial fire seven times. The bride goes to live in her husband's house for three days and then returns to her own house to await puberty, when she will be reunited with her husband.

Italians

After the wedding ceremony, Italian newlyweds are showered with confetti made of sugar-coated almonds. This confetti symbolizes the bitterness and sweetness of married life.

Japanese

Japanese couples are traditionally introduced by a *nakodo*, or go-between, who is usually a friend or relative. The engagement is celebrated with a toast of sake (Japanese wine made from rice) and an exchange of presents such as seaweed, fish, fans, and thread. The most common wedding ceremony in Japan is the Shinto ceremony. The bride and groom sit at the altar of a shrine with their parents and the nakodo. After being purified by a Shinto priest, the bride and groom each drink from three cups of sake three times. The bride wears a white kimono to symbolize the death of her ties to her own family. She also wears a special hat known as a "horn cover" to cover her horns of jealousy. The marriage is legal when the couple registers at a local government office.

Mbutis

These nomadic people live in central Africa. A Mbuti man must prove his worth to a woman's parents by catching an antelope single-handedly and offering it to them. He also gives small gifts of roots, nuts, or birds, or orchids from the

DID YOU KNOW?

Marrying by the Millions

There are about 2.3 million marriages performed in the United States every year. That breaks down to about 6,400 every day.

tops of the tallest trees in the forest. When the couple is ready to be married they build a house and live together. They are finally married three days after the bride gives birth to her first child.

FIVE LOVE STORIES TO KNOW

Antony and Cleopatra

One of the most famous women in history, Cleopatra was the brilliant and beautiful last pharaoh of Egypt. When Roman general Marc Antony went to Egypt to advance the growing power of Rome, he fell in love with Cleopatra. Their affair scandalized Roman society and bothered Roman politicians, who were suspicious of Egypt's power. Yet despite the risks, Antony and Cleopatra married and planned to conquer Rome. During a battle, Antony heard a false report that Cleopatra was dead and, devastated, fell on his sword. With no hope for a grand future left, Cleopatra induced a poisonous asp to bite her and thus died.

Shah Jahan and Mumtaz Mahal

In 1612, a teenage girl, Arjumand Banu, married fifteen-year-old Shah Jahan, ruler of India's Mughal Empire. Renamed Mumtaz Mahal, she bore Shah Jahan fourteen children and became his favorite wife. After Mumtaz died in 1629, the grieving emperor resolved to create a fitting monument. It took twenty thousand workers and one thousand elephants nearly twenty years to complete this monument — the Taj Mahal. Many people consider the white marble tomb, topped with a dome and surrounded by four minarets, to be the most beautiful building in the world.

Abigail and John Adams

Although she lived at a time when few women were educated, Abigail Smith learned to read and developed an appreciation of current events. Her lively mind caught the attention of a young lawyer, John Adams, and they were married in 1764. It was an intellectual and romantic relationship that would last for more than fifty years. The Revolutionary War and John's various diplomatic missions often forced them to be apart for long periods of time, so they wrote each other long, affectionate letters that are now famous. When she did join her husband, on diplomatic missions in Europe or during his presidency, Abigail was a valued partner, entertaining with style and offering shrewd observations.

Queen Victoria and Prince Albert

Victoria ascended the throne of England in 1837. Three years later, she married her first cousin, Prince Albert. The couple had nine children and loved each other deeply. When Albert died in 1861, Victoria was devastated. She did not appear in public for three years. Victoria never stopped mourning her beloved prince, wearing black until her death in 1901. During her reign, the longest in British history, Britain became a world empire on which "the sun never set."

Annie Oakley and Frank Butler

In 1881, the famous Baughman and Butler shooting act was performing in Cincinnati. The star of the show, champion shot Frank E. Butler, boasted that he could beat any local marksman. Butler was amused when told that a young woman had accepted his challenge. But Annie Oakley won the contest. She also captured Butler's heart, and they were married the next year. Butler abandoned his career to manage hers, which brought her fame in Buffalo Bill's Wild West Show. Their happy marriage lasted forty-four years, and their love story was immortalized in the musical *Annie Get Your Gun*.

WEDDING LORE AND TRADITIONS

Have you ever wondered why a bride tosses her bouquet, or why the wedding ring is worn on the third finger of the left hand? The origins of and meaning behind some of our most cherished wedding traditions may surprise you. There are, of course, multiple explanations for each piece of wedding lore, and few customs can be definitively traced back to their roots. Below are some of the more common stories behind these traditions.

Tossing the Bouquet

Tossing the bouquet is a tradition that stems from England. Women used to try to rip pieces from the bride's dress and flowers in order to obtain some of her good luck. To escape from the crowd, the bride would toss her bouquet and run away. Today the bouquet is tossed to single women with the belief that whoever catches it will be the next to marry.

Giving Away the Bride

The tradition of the bride's father giving away his daughter has its roots in the days of arranged marriages. Daughters in those times were considered their fathers' property. It was the father's right to give his child to the groom, usually for a price. Today a father giving away his daughter is a symbol of his blessing of the marriage.

The Wedding Ring

The wedding ring has been worn on the third finger of the

"Love doesn't sit there, like a stone. It has to be made, like bread; remade all the time, made new."

— Ursula Le Guin, American writer

"When you love someone, all your saved-up wishes start coming out."

— Elizabeth Bowen, Irish novelist

"I love being married. It's so great to find the one special person you want to annoy for the rest of your life."

— Rita Rudner, American comedienne

"A difference in taste in jokes is a great strain on the affections."

— George Eliot (Mary Ann Evans), British novelist

"How do I love thee? Let me count the ways. / I love thee to the depth and breadth and height / my soul can reach. . . ."

— Elizabeth Barrett Browning, British poet

left hand since Roman times. The Romans believed that the vein in that finger ran directly to the heart. The wedding ring is a never-ending circle, which symbolizes everlasting love.

Something Old, Something New, Something Borrowed, Something Blue

"Something old" represents the bride's link to her family and the past. The bride may choose to wear a piece of family jewelry or her mother's or grandmother's wedding gown.

"Something new" represents hope for good fortune and success in the future. The bride often chooses the wedding gown as the new item.

"Something borrowed" usually comes from a happily married woman and is thought to lend some of her good fortune and joy to the new bride.

"Something blue" is a symbol of love, fidelity, and the purity of the bride.

Body and Mind

LANDMARKS

1832 American educator Catherine Beecher publishes the first book about physical fitness for women, *A Course of Calisthenics for Young Ladies.*

1890s Bicycling becomes a health trend for women. It's not just the wheels that enable women to move more quickly and freely — bicycling helps spark the popularity of shorter skirts and looser, more comfortable clothing.

1920 For the first time, menstruating women can use disposable cotton pads, called Kotex, which were developed from bandages used during World War I. Now women can just throw out their pads instead of having to wash and reuse cotton rags.

1958 Two years after its founders first met, La Leche League is officially founded to educate women about breast-feeding and its health benefits. During the 1950s and 1960s, breast-feeding went out of fashion in the United States.

1972 Rates of breast-feeding in the United States begin to climb again. Over the next twenty years, they will nearly double.

1973 The Boston Women's Health Book Collective publishes the first edition of their groundbreaking book, *Our Bodies, Ourselves.*

1974 The U.S. government makes the Special Supplemental Program for Women, Infants, and Children (WIC) permanent. WIC helps provide food supplements and nutritional counseling to pregnant women and new mothers — more than seven million per month in the year 2001.

FEMALE EVENTS

The female body is incredible. The following list describes the events in a female's body that enable her to bear children.

At birth, a baby girl has about four hundred thousand immature eggs, or ova, in her ovaries.

During puberty the eggs begin to mature. Each month one egg ripens and leaves the ovary. It passes through a fallopian tube, where, if not fertilized by a sperm, it disintegrates. The uterus, which has built up tissue and blood to make a nourishing nest for a fertilized egg, sheds its lining about a week after the egg disintegrates. This is the event known as menstruation, or the menstrual period.

Gestation, or pregnancy, begins when an egg that has been released from the ovary is fertilized by a sperm. The result is the eventual birth of a child.

Lactation is the production of milk in a woman's breasts to feed her newborn child.

Menopause is when a woman's ovaries gradually stop functioning. Menstruation ceases, and this marks the end of her ability to bear children.

FROM GIRL TO WOMAN:
HOW YOUR BODY CHANGES

The passage from girlhood to womanhood is called puberty. It begins for girls anywhere from age eight to sixteen; the average age is eleven. What happens to your body during puberty?

You experience a growth spurt. Sometimes your arms, hands, legs, and feet may seem to grow faster than the rest of your body. (But don't worry — that clumsy feeling will pass.)

Your internal reproductive system matures: your ovaries and uterus grow larger.

Secondary sex characteristics appear: budding breasts and pubic hair. It's common for one breast to begin developing before the other.

The shape of your body begins to change, typically becoming softer and more rounded. Your fat-to-muscle ratio increases. You will probably notice some more fat around your hips, bottom, stomach, and legs.

Hormonal changes bring an increase in perspiration and oily skin. That's why pimples can start occurring at this time.

You begin your menstrual periods. The first menstrual periods are often light and irregular. You may menstruate once, then not again for several months. Once they become regular, a menstrual period occurs every twenty-five to thirty-five days (the average is twenty-eight days). Bleeding lasts three to seven days. Menstruation is a sign that you are able to become pregnant.

Female Reproductive System: Internal

Ovaries:
The organs that hold the ova, or eggs, and where the ova ripen.

Fallopian tubes:
The passageways from the ovaries to the uterus traveled by the ripened ova.

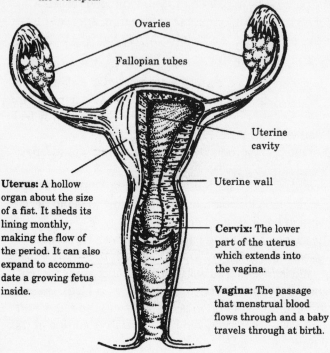

Ovaries

Fallopian tubes

Uterine cavity

Uterine wall

Uterus: A hollow organ about the size of a fist. It sheds its lining monthly, making the flow of the period. It can also expand to accommodate a growing fetus inside.

Cervix: The lower part of the uterus which extends into the vagina.

Vagina: The passage that menstrual blood flows through and a baby travels through at birth.

SOME OTHER TERMS FOR MENSTRUATION

Girls and women have come up with all sorts of creative terms for their periods. Sometimes they have done so because they think that menstruation is embarrassing, but sometimes it's just because "menstruation" is such a big, clinical word! Here are some of the more common nicknames.

"Aunt Flo is visiting."

"Having my cycle."

"Big Red."

"Decorated with red roses."

"The painters are here."

"It's that time of the month."

"My calendar days."

"Monthlies."

"Bloody Mary."

"Moon days."

"It's a red-letter day!"

"Flying the red flag."

Female Reproductive System: External

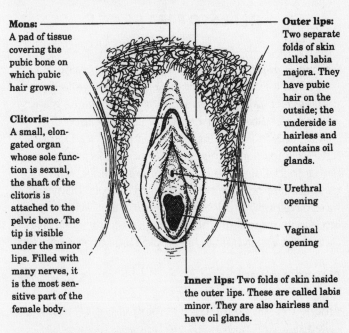

Mons: A pad of tissue covering the pubic bone on which pubic hair grows.

Clitoris: A small, elongated organ whose sole function is sexual, the shaft of the clitoris is attached to the pelvic bone. The tip is visible under the minor lips. Filled with many nerves, it is the most sensitive part of the female body.

Outer lips: Two separate folds of skin called labia majora. They have pubic hair on the outside; the underside is hairless and contains oil glands.

Urethral opening

Vaginal opening

Inner lips: Two folds of skin inside the outer lips. These are called labia minor. They are also hairless and have oil glands.

Aunt Flo Is Visiting Earlier

In the United States today, girls begin menstruating earlier than they once did, mostly because of better nutrition. Around the time of the Civil War, the average age for a girl's first period was seventeen. One hundred years later, the average age had dropped to about twelve and a half. That age has remained the same since about the 1950s.

RITES OF PASSAGE

Getting your period is a sign of maturity or coming of age — a cause for celebration! In America, coming-of-age ceremonies celebrating menstruation are rare because many girls are raised to think that they should talk about their periods only in private. Recently, some American families have tried to change this by honoring a girl's first period with a special dinner, a card of congratulations, or even a candlelit ceremony. But for hundreds of years, many cultures have publicly celebrated menstruation, acknowledging it as an important sign of female maturity and fertility — without which human life would come to an end!

Native America

Among the Navajo there is a coming-of-age ceremony called *kinaalda*. Young girls run footraces to show their strength. They also bake a huge cornmeal pudding for the whole community to taste. During the ceremony girls wear special clothing and arrange their hair to imitate the goddess Changing Woman.

The Nootka Indians of the Pacific Northwest consider a girl's first period a time to test her physical endurance. She is taken way out to sea and left there. She must then swim back to shore, where she is greeted and cheered by the whole village.

The Mescalero Apaches consider the female puberty ceremony a most important celebration in their tribe. Each year an eight-day event honors all girls who started their periods that year. It begins with four days of feasting and dancing. Boy singers recount the tribe's history with songs each night. A four-day private ceremony follows, in which the girls reflect on the changes in their bodies and their passage into womanhood.

Australia

Among the Aborigines of Australia, a girl is treated to the tradition of "love magic" when she gets her first period. The women of the tribe sing and teach her about the female powers and the physical changes that mark womanhood.

Japan

When a Japanese girl gets her first period, her family throws a big party. Family and friends are invited but are not told why they are celebrating. When the girl's family brings out a tray bearing a decorated pear, a candied apple, or red-colored rice and beans, the secret reason for the party is revealed.

Micronesia

The Ulithi tribe call a girl's rite of passage *kufar*. When a girl begins her first period, she must go to a menstrual house. She is joined by women of the village, who bathe her and recite magic spells. She will always return to the menstrual house during her period.

Nigeria

The Tiv tribe literally mark a girl at the time of her first period. Four lines are cut into her abdomen. The resulting scars represent her womanhood and are thought to make her more fertile.

Sri Lanka

When a girl gets her first period in Sri Lanka, the exact time and day are noted. An astrologer is consulted to make predictions about the girl's future based on the alignment of the stars at that time. The family then prepares the house for a ritual bathing, in which the women of the family wash the girl's hair and scrub her all over. She then puts on new white clothes, from her underwear to her shoes. Printed invitations are sent for a party at which gifts and money are presented to the girl.

BODY TRAPS

Being female has never been easy, especially when you consider the traps and tortures inflicted on girls and women for the sake of someone's idea of beauty. Here's a look at some of the life-threatening ways girls and women have tried to reshape themselves.

Feet

In China, beginning in the eleventh century, the practice of foot binding was used on female infants in wealthy families. The feet were tightly bound to prevent growth. The toes sometimes dropped off, and a deep cleft formed between the heel and the front of the foot. These unnaturally small feet were considered a sign of beauty. Foot binding was intended to distinguish natural-footed working women from wealthy women of leisure, whose bound feet confined them to the house because walking was so painful. This practice was outlawed in the twentieth century.

Waists

In the late 1800s, some women had their two lowest ribs surgically removed. This way, with the help of a corset, they could achieve an even smaller waist than women who had a normal number of ribs.

Breasts

In the 1960s, some American women had liquid silicone injected into their breasts to enlarge them. This dangerous procedure often backfired. The silicone sometimes solidified and traveled through the body, causing infections and odd lumps in surprising places.

Today, some women have saline-filled pouches surgically implanted beneath their natural breasts to push them forward. The risks of this kind of surgery include pain, infection, leaking implants, a buildup of scar tissue, and the possibility that an implant may tighten up so that the breast appears deformed. Also, implants do not last forever; many rupture or deflate within a few years, or sometimes just a few months, making additional surgery necessary.

Necks

Among the Padaung women of Burma, long necks are signs of beauty. Young girls wear brass or iron rings around their necks in order to stretch them. Beginning at five, the number of rings increases to a total of twenty-two in adulthood. The bones of the neck are pulled apart, and the rings can never be removed without the risk of death. These "beautiful" necks are stretched to lengths of fourteen inches.

Lips

In Africa, girls of the Sras Djunge tribe begin to stretch their lips by implanting wooden disks in their mouths at the age of four. As the girls grow, they use larger disks, and their lips stretch so far that they are barely able to talk and can consume only liquids.

Body Weight

Today, many American girls and women starve themselves in the name of beauty. Many people believe that an obsession with body weight, especially in the media, is a key part of why so many girls develop eating disorders. There are a number of different kinds of eating disorders. Anorexia nervosa is an eating disorder characterized by self-starvation. Bulimia is an eating disorder in which a person eats but later throws up the food. All eating disorders include unhealthy attitudes toward food and a poor self-image.

And all types of eating disorders are dangerous. The dangers include malnutrition, dental problems, heart problems, an increased risk for such diseases as diabetes and arthritis, and even death. Thousands of girls and women die of eating disorders each year. If your own feelings about food and your body have begun to worry you, or if you are worried about a friend, tell someone you trust. Reaching out is the first step toward health.

FOOD FOR THOUGHT

- Fat was once called "the silken layer."

- The Victorians associated plumpness with health, attractiveness, and a happy outlook.

"The body says what words cannot."
— Martha Graham, American dancer
and choreographer

"For fast-acting relief, try slowing down."
— Lily Tomlin, American comedienne

"I just told my mother I want a bra. Please help me grow, God. You know where."
— Judy Blume, in *Are You There, God? It's Me, Margaret*

"Whenever I get my period (and that's only been three times), I have the feeling that in spite of all the pain, discomfort and mess, I'm carrying around a sweet secret."
— Anne Frank, in *The Diary of a Young Girl*

"Just imagine what would happen if we were to take all the energy we expend trying to conform to society's standard of beauty and direct it somewhere else."
— The Boston Women's Health Book Collective, in *Our Bodies, Ourselves for the New Century*

- Fat is a sign of fertility. The soft roundness of a woman's hips, thighs, belly, and breasts is a sign that she is a fertile adult.

- This soft female curviness has been considered attractive and desirable in most cultures throughout most of human history.

- Ultraskinny models weigh 25 percent less than the average American woman.

- Models in magazines can look very different than they do in real life. In magazines, lighting, makeup, photograph retouching, and other special effects contribute to an unreal look.

- Surveys show that women tend to be much more critical of women's bodies than men are.

- If Barbie the doll were a human, she would probably have to crawl on all fours, because her tiny feet could not support her long legs and oversized chest.

- Real bodies come in every shape and size.

GOOD-FOR-YOUR-BODY FOODS

Folklore has long taught that the following foods are good for you. Now scientists agree.

Cranberry juice is good for urinary tract health. Why? Because the juice inhibits a type of bacteria that clings to the wall of the bladder and causes infection.

Carrots are good for your eyes. Carrots and some other fruits and vegetables contain beta carotene, which can reduce the chance of eye disease. One carrot a day can help prevent macular degeneration, which eventually leads to blindness.

Chicken soup fights the congestion that comes with a cold. Chicken has an amino acid that thins the mucous lining of the sinuses, thus relieving stuffiness.

Garlic and **onions** kill flu and cold viruses.

Fish is good for your brain. The mineral zinc is found in fish and shellfish. Studies show that even a minimal deficiency of zinc impairs thinking and memory.

Blueberries fight the bacteria that causes diarrhea.

Bananas are a natural antacid. They soothe heartburn or gastric distress.

Spinach is good for your spirits. It contains lots of folic acid. If your body doesn't have enough folic acid, you may feel depressed.

Ginger root fights the nausea caused by motion sickness and relieves migraine headaches. Make a tea of fresh ginger root by simmering it in water for ten minutes.

Nuts are packed with vitamins and minerals that help to boost brain function and lower cholesterol.

Onions can fight insomnia. Onions contain a mild natural sedative called quercetin.

Yogurt with acidophilus fights the bacteria that causes vaginal yeast infections.

EXERCISE — IT'S GOOD FOR YOU!

Exercise is not only fun; it is also good for your body, mind, and overall well-being. Kids who exercise on a regular basis often do better in school, sleep better, don't feel as tired, are less likely to get hurt while exercising, and are stronger than less active children. Exercise also helps to relieve stress and anxiety.

There are two types of exercise, **aerobic** and **anaerobic.** When you do an aerobic exercise, such as walking, running, swimming, inline skating, jumping rope, playing soccer, dancing, bicycling, or playing hockey, you increase your heart rate and the flow of oxygen-rich blood to the muscles. Aerobic exercise builds endurance and burns fat and calories. Doctors recommend that people do thirty minutes of aerobic exercise every day. When you do an anaerobic exercise, such as weightlifting or push-ups, the short bursts of exertion build strength and muscle mass.

FEELING GOOD

Your body and your mind work together — one influences the other. Here are some of the things you can do with your body and mind to feel good about yourself.

• Put on comfortable clothes and go out for a walk. Walking helps clear your head so that you can collect your thoughts.

• Take one evening a week just to chill out. Snooze, read a book, or play a game with friends.

• Take a warm, soothing herbal bath: combine bay leaves and thyme and tie them in a washcloth. Throw it into the running bath water with a cup of salt.

• Write in a diary at least twice a week. Keep track of how you feel. Don't hold anything back.

• Listen to your favorite music.

• Paint a mural or self-portrait.

• Keep a journal of your dreams.

• Try to relax all the muscles in your body by first tensing them and then letting them go. Work from your toes on up to your head.

• Practice yoga or stretching — they are great stress reducers and body toners. Yoga promotes strength, balance, flexibility, good posture, and a general sense of well-being.

DID YOU KNOW?

Body Knowledge Is Powerful

In 1969, twelve women who met at a feminist conference in Boston formed a discussion group about their bodies and health. These women found that they shared many experiences. They had been to doctors who had talked down to them, ignored their concerns, and withheld information. They discovered that, in many ways, they understood their bodies better than their doctors did. They also discovered that learning about their health felt powerful.

To learn more — and to share what they learned with other women — the group did research, wrote papers, and conducted interviews with health experts. In 1972, they organized as the Boston Women's Health Book Collective, and in 1973, they published the first edition of a landmark book, *Our Bodies, Ourselves.* The book has helped millions of women to "become their own health experts" by learning about their own bodies as well as the experiences of other women. Since 1980, the collective has also published a health book for teens called *Changing Bodies, Changing Lives.*

- To take your mind off problems, find a quiet spot, close your eyes, relax your body, and picture yourself in perfectly peaceful surroundings.

- When you are stressed or emotionally jammed up, close your eyes and take three very deep breaths, each time exhaling completely until most of the air is out of your lungs. This will refresh you and help you to think more clearly.

- Dance! You can twist and shake by yourself or find a broom to be your waltz partner.

BLUE

You might wake up one morning and just feel like saying, "Blah." Maybe it's raining outside. Or maybe you have to take a quiz you didn't study for. Or maybe you had a fight with your brother or sister the night before, and you said mean things to each other. Or maybe you don't know why you feel bad.

Feeling blue, or depressed, can happen to kids and adults. It usually lasts a day or two and then passes. If you keep feeling blue for days and days, try to talk to someone about your feelings. Parents, teachers, school counselors, and religious leaders are all people who can help you.

FIVE WOMEN TO KNOW

Writer and educator **Catherine Beecher** (1800–1878) was a pioneer in health and physical education for American girls and women. She developed a widely used system of calisthenics that was designed to make women's bodies stronger and healthier. Beecher also spoke out against the wearing of corsets, which she believed were hazardous to internal organs, and wrote extensively about the benefits of good nutrition.

Senda Berenson (1868–1954) developed the game of women's basketball, which she introduced at Smith College in 1893. The game was a landmark for women's athletics and quickly gained popularity at schools across the United States. Like the recently invented men's basketball, it offered an effective means of indoor exercise, but Berenson amended the rules for women to avoid unladylike "rough play."

With the publication of her book *Diet for a Small Planet* in 1971, **Frances Moore Lappé** (b. 1944) helped millions of people to learn about eating in a more healthy way. Her work is still important because it ties together a number of health concerns, from nutrition to junk food advertising to world hunger. Many vegetarians say Lappé's book helped get them started with a meatless diet.

Dr. Susan Love (b. 1948), a surgeon and activist, is a leader in breast cancer research and advocacy. She has helped to raise awareness about breast cancer and to educate women about breast health. She cofounded the National Breast Cancer Coalition and wrote the bestseller *Dr. Susan Love's Breast Book.*

In 1990, **Dr. Antonia Novello** (b. 1944) became the first woman to serve as U.S. surgeon general. She focused on publicizing the dangers of smoking and teenage drinking, expanding AIDS education, and improving health care for women, children, and people of color. In 1993, she left her post as surgeon general to work for the United Nations Children's Fund (UNICEF).

Sports

When the first Olympic Games were held in ancient Greece, women were not even permitted to view the events. Today women are making and breaking records in the Olympics and in many sports. Here are some of the sports in which women have excelled, and some landmarks and history makers in each.

BASEBALL

Women have been playing baseball since 1866. Vassar College had the first women's baseball team.

The first woman ever to sign a professional baseball contract was Lizzie Arlington, in 1898. The twenty-year-old pitched one game for the Reading, Pennsylvania, Class A Atlantic League.

Amanda Clement was the first official female umpire. She umpired from 1905 to 1911 for Midwestern semiprofessional teams. She designed her own uniform — an ankle-length skirt, a white shirt with a black tie, and a baseball cap. She stored extra baseballs in her blouse.

The All-American Girls Professional Baseball League, or AAGPBL, debuted in 1943 to give baseball fans something to watch during World War II, when male players were away fighting the war. Philip Wrigley, a chewing gum mogul and owner of the Chicago Cubs, started the league and soon found that there was an abundance of women with baseball talent in the United States and Canada. The league enjoyed years of success before eventually folding in 1954. *A League of Their Own,* a movie starring Geena Davis, Rosie O'Donnell, and Madonna, depicted the early years of the league.

Girls were officially admitted to Little League Baseball on June 12, 1974.

In 1984, Victoria Roche became the first girl to play in the

Little League World Series, in Williamsport, Pennsylvania.
Ila Borders was the first woman to earn a scholarship for
baseball and the first to win a men's college game, with
the Southern California College Vanguards in Costa
Mesa, California, in 1994. She also was the first woman
to pitch in a regular-season minor league baseball game
when she joined the St. Paul Saints in 1997.

BASKETBALL

The first women's college basketball game was played at
Smith College in 1893.
The Women's Basketball Association, a professional basket-
ball league, was founded in 1977. The WBA started with
eight teams and lasted three seasons.
In 1977, Lucy Harris became the first woman to be drafted
by an NBA team (the New Orleans Jazz). In 1979, Anne
Meyers signed an NBA contract for $50,000 for one
year with the Indiana Pacers. Neither ever appeared in a
game.
The Harlem Globetrotters are the only coed professional
basketball team. In 1985, the team picked Olympian
Lynette Woodward as its first female member.
Nancy Lieberman, an outstanding basketball player from
Old Dominion College in Virginia, played for Spring-
field, Massachusetts, in the U.S. Basketball League in
1986, becoming the first woman in history to play in
a men's professional league.
The Women's National Basketball Association (WNBA) was
established in 1997. The Houston Comets, led by superstars Cynthia Cooper and Sheryl Swoopes, won the
league championship its first four years, from
1997–2000.
There are sixteen WNBA teams — the Charlotte Sting,

Cleveland Rockers, Detroit Shock, Houston Comets, Indiana Fever, Los Angeles Sparks, Miami Sol, Minnesota Lynx, New York Liberty, Orlando Miracle, Portland Fire, Phoenix Mercury, Seattle Storm, Sacramento Monarchs, Utah Starzz, and Washington Mystics.

FIELD HOCKEY

Constance M. K. Applebee is credited with bringing field hockey to the United States in 1901. She had come from England to the United States to take a physical education course at Harvard University. The first game she organized was played on concrete with ice hockey sticks in a courtyard at Harvard.

In 1922, women formed the United States Field Hockey Association. Today the association has more than fifteen thousand members.

Field hockey was added to the Olympics in 1980. The Zimbabwe women's field hockey team went undefeated in the tournament that year, winning the sport's first Olympic gold medal. In 1996, Australia took the gold medal, becoming the first country to win two golds in women's field hockey. At the 2000 Sydney Games, the Aussies defended their title on their home turf for a third gold.

FOOTBALL

Pat Palinkas was the first woman to play in a professional football game. In 1970, she held the ball for the place kickers on the Orlando Panthers team.

On December 27, 1987, Gayle Sierens became the first woman to do play-by-play commentary for the broadcast of a National Football League game, between Kansas City and Seattle.

In 2000, the Women's Professional Football League (WPFL) kicked off its inaugural season. The WPFL ball is slightly smaller than the one used in the NFL, but the rules are the same. Eleven teams took part in the league in its first year — the Austin Rage, Colorado Valkyries, Daytona Beach Barracudas, Houston Energy, Miami Fury, Minnesota Vixens, New England Storm, New York Galaxy, New York Sharks, Oklahoma City Wildcats, and Tampa Tempest. In addition, there were two exhibition teams — the Carolina Cougars and Sacramento Sirens.

GOLF

Mary, Queen of Scots, was probably the first woman to play golf. It was during her reign that the famous golf course at St. Andrews, Scotland, was built, in 1552. Mary coined the term "caddy" by calling her assistants cadets.

Patty Berg won the first U.S. Women's Open in 1946.

The Ladies Professional Golf Association's player of the year from 1966 to 1973 (with the exception of 1970) was Kathy Whitworth of Texas. During her career, which lasted from 1962 to 1985, she won eighty-eight championships, more than any other professional female golfer. She was inducted into the LPGA Hall of Fame in 1975.

Nancy Lopez was the first female golfer to be rookie of the year and player of the year in the same year. She won both awards in 1978.

In 1990, Juli Inkster of Los Altos, California, became the

first woman to win the only professional golf tournament in which women and men compete head-to-head. She won the Invitational Pro-Am at Pebble Beach, California, in a one-stroke victory.

There are now approximately forty LPGA tournaments played each year, including four "major" tournaments. They are the Nabisco Championship, the U.S. Women's Open, the LPGA Championship, and the Women's British Open.

Australian Karrie Webb is the youngest woman to win a Career Grand Slam, winning six majors in all as of 2002. Her win in the 2002 Women's British Open made her the first player to win the "Super Grand Slam" (five different majors).

GYMNASTICS

The first women's gymnastics instruction in the United States was given at Mount Holyoke College in 1862.

Larissa Latynina of Russia won eighteen Olympic gymnastics medals, setting an Olympic record for women. She won nine gold medals, five silver, and four bronze between 1956 and 1964.

Marcia Frederick was the first American woman to win the world gymnastics title. She won in 1978 on the uneven bars.

In the 1976 Olympics, Nadia Comaneci of Romania became the first gymnast in Olympic history to score a perfect 10. She did this in the floor exercise. She went on to record six more perfect scores during those Olympics.

In 1984, sixteen-year-old Mary Lou Retton of West Virginia became the youngest gymnast to win an Olympic gold medal. Her perfect performance on the vault earned her the gold in the women's all-around event.

Shannon Miller is now the most decorated female American gymnast of all time. She has seven Olympic medals altogether — two gold, two silver, and three bronze.

Kerri Strug provided one of the most exciting stories of the 1996 Olympics and gymnastics history when she nailed her vault on an injured foot to ensure the gold medal for the United States team. Strug was just one member of the gold medal–winning team known as "the Magnificent Seven." The others were Amanda Borden, Amy Chow, Dominique Dawes, Shannon Miller, Dominique Moceanu, and Jaycee Phelps.

HORSE RACING

The first female jockey was Alicia Meynell of England. She first competed in a four-mile race in York, England, in 1804.

In 1934, Mary Hirsch became the first licensed female trainer of thoroughbred race horses.

In 1970, Diane Crump became the first female jockey to ride in the Kentucky Derby, the biggest thoroughbred race of the year.

In 1991, Julie Krone became the first female jockey to ride in the Belmont Stakes, which, like the Kentucky Derby, is part of horse racing's Triple Crown. She retired in 1999 with more wins than any other female jockey in history, and as one of the most successful jockeys of all time.

ICE HOCKEY

On September 23, 1992, Manon Rheaume started as goalie for the Tampa Bay Lightning in a National Hockey

League exhibition game. She was the first female to start a game in any of North America's four major professional sports leagues (baseball, football, basketball, and ice hockey).

In each of the first seven Women's World Hockey Championship tournaments (1990, 1992, 1994, 1997, 1999, 2000, and 2001), Canada won the gold and the United States won the silver.

Women's hockey was accepted as a full medal sport at the 1998 Olympic Games in Nagano, Japan. The United States was undefeated (6–0) in the tournament that year and won the gold medal with a thrilling 3–1 victory over Canada.

Women's ice hockey quickly gained popularity after the 1998 Olympics. Many colleges and universities offered women's ice hockey, and finally, in 2000, it became a sanctioned NCAA (National Collegiate Athletic Association) championship sport.

RUNNING

Long Distance

In 1966, Roberta Gibb became the first woman ever to run in the famed Boston Marathon when she entered as a man. She wore a hooded sweatshirt to cover her face and hair. Unofficially she finished in 125th place out of approximately 500 entrants with a time of 3:21:25.

In 1967, Kathrine Switzer also ran the Boston Marathon. Because it was still a male-only event, she registered as K. Switzer and ran the entire route with officials attempting to tear her number from her back. She estimated her time at just over four hours and twenty minutes. Her run created such a stir that the AAU, the

Amateur Athletic Union of the United States, rallied to get the rules changed. In 1972, after a long and hard five-year battle, Switzer became one of the first nine women to run officially and legally in the Boston Marathon.

Grete Waitz of Norway has won the women's race in the New York City Marathon nine times, more than any other competitor. She won in 1978, 1979, 1980, 1982, 1983, 1984, 1985, 1986, and 1988.

Kenya's Catherine Ndereba, the 2000 and 2001 Boston Marathon female winner, ran the fastest women's marathon ever when she won the 2001 Chicago Marathon in 2 hours, 18 minutes, and 47 seconds. The American record is held by Olympic champion Joan Benoit Samuelson, who ran a 2:21:21 in Chicago in 1985.

Short Distance

In 1932, Babe Didrikson Zaharias became the only athlete, man or woman, to win an Olympic medal in a running event (80-meter hurdles), a throwing event (javelin), and a field event (high jump).

Chi Cheng of Taiwan was the first woman to run 100 yards in 10 seconds flat. She accomplished this on June 13, 1970, in Portland, Oregon.

Wilma Rudolph was the first woman ever to win three track-and-field gold medals at one Olympic Games. She earned them for the 100-, 200-, and 400-meter races in 1960.

Florence Griffith Joyner, nicknamed "Flo-Jo," became the fastest woman in the world when at the 1988 Olympics she ran the 200-meter race in 21.34 seconds.

Jackie Joyner-Kersee was given the title of "World's Greatest Athlete" by *Track & Field News* and owns three gold medals, as well as the world heptathlon record.

Evelyn Ashford has more Olympic track-and-field gold medals than any other American woman. She has four golds to go along with her one silver.

American superstar Marion Jones is the best female sprinter and one of the best female long jumpers in the world. Jones, a former basketball star at the University of North Carolina, was attempting to win five gold medals at the 2000 Olympic Games in Sydney. She came away with three golds (in the 100-, 200-, and 4x400-meter relays)

SHE SAID IT

"All of my life I have always had the urge to do things better than anybody else."

— Babe Didrikson Zaharias, American all-around star athlete

"Your opponent, in the end, is never really the player on the other side of the net, or the swimmer in the next lane, or the team on the other side of the field, or even the bar you must high-jump. Your opponent is yourself, your negative internal voices, your level of determination."

— Grace Lichtenstein, American writer

"The first thing is to love your sport. Never do it to please someone else. It has to be yours."

— Peggy Fleming, American figure skater

"A horse doesn't know whether the rider on his back wears a dress or pants away from the track."

— Diane Crump, American jockey

"Jumping has always been the thing to me. It's like leaping for joy."

— Jackie Joyner-Kersee, American track-and-field star

and two bronzes (in the long jump and the 4x100-meter relay).

SKIING

Skiing has been an Olympic event for women since 1936, when Germany's Christl Cranz won the first women's alpine skiing gold medal, in the alpine combined.

The first American woman to win an Olympic skiing gold medal was Gretchen Fraser, who won the gold in the slalom in 1948. In 1998, Picabo Street came through for the United States by winning the gold in the super giant slalom.

The alpine skiing World Cup competition has taken place every year since 1967 in various locations. Switzerland and Austria usually dominate the women's competitions, although Germany's Katja Seizinger won in 1996 and 1998. Tamara McKinney won the World Cup for the United States in 1983, and Canada's Nancy Greene won the first two, in 1967 and 1968.

Today there are more than thirty women's World Cup skiing events that take place all over the world.

SOCCER

In 1991, Jo Ann Fairbanks became the first American female referee to serve at an international soccer event when she was a lineswoman in the women's qualifying rounds for the North and Central American and Caribbean regional soccer tournament in Haiti.

The first women's World Cup soccer championship was won by the United States in December 1991.

Women's soccer became an Olympic sport in 1996, and the U.S. team won the gold medal. The Americans also took home the silver in Sydney in 2000, losing in overtime in the gold-medal game to Norway, 3–2.

American Mia Hamm became the all-time leading female scorer in international soccer with her 108th goal on May 22, 1999.

In front of 90,185 fans (the largest crowd ever to watch a women's sporting event) at the Rose Bowl in Pasadena, California, the U.S. women's national soccer team played China to a 0–0 tie in the final match of the 1999 World Cup. The United States won the game and its second World Cup by beating China 5–4 in a penalty kick shootout. Brandi Chastain scored the game winner for the Americans.

A new American women's professional league, the Women's United Soccer Association (WUSA), made its debut in 2001. Eight teams competed in the inaugural season — the Atlanta Beat, Bay Area CyberRays, Boston Breakers, Carolina Courage, New York Power, Philadelphia Charge, San Diego Spirit, and Washington Freedom. In the 2001 championship game, Bay Area defeated Atlanta on penalty kicks (4–2) after the two teams battled to a 3–3 tie in regulation.

SOFTBALL

The first world softball championship was a women's tournament held Melbourne, Australia, in 1965.

Joan Joyce was a legendary softball pitcher. During twenty years of competition (1955–1975), she won 509 games and lost only 33.

In 1982, softball became an officially sanctioned NCAA (National Collegiate Athletic Association) sport. With eight national championships, UCLA has had the most successful Division I program to date.

In 1995, the Women's Professional Fastpitch Tour traveled around the United States to bolster interest in the new fastpitch women's league. In November 2002, the name of the league was changed to National Pro Fastpitch (NPF). The league consists of four teams — the Akron Racers, Florida Wahoos, Ohio Pride, and Tampa Bay FireStix.

Softball became an official Olympic sport in 1996. The United States won the gold medal that year, while China took the silver. The United States defended its title in Sydney in 2000, edging previously undefeated Japan 2–1 in the gold-medal game.

SWIMMING

Swimming became an Olympic event in 1908, but women weren't allowed to compete until 1912. Fanny Durack of Australia became the first woman to win a gold medal in the 100-yard freestyle race that year.

Gertrude Ederle was the first woman to swim the English Channel. In 1926, she swam from France to England in 14 hours and 39 minutes.

Florence Chadwick was the first woman to swim both ways across the English Channel. In 1950, she swam from France to England in 13 hours, 20 minutes. She swam the other direction, from England to France, in 1951, 1953, and 1955.

Janet Evans won four gold medals in the 1988 and 1992 Olympics. Amy Van Dyken won four in the 1996 Olympics and then added two more in 2000. But it's

Jenny Thompson who holds the Olympic record for female swimmers, with eight gold medals between 1988 and 2000. She also has one silver and one bronze, giving her ten total, more than any other female swimmer in the world.

Marathon swimmer Susie Maroney of Australia swam from Isla Mujeres, Mexico, to Las Tumbas, Cuba, in 1997. It was a world-record swim of 128 miles that took her 38 hours and 27 minutes to complete.

TENNIS

Mary Ewing Outerbride is credited with introducing lawn tennis to the United States in 1874.

The first American woman to win the women's singles title at Wimbledon was May Sutton Bundy. She won in 1904 and in 1907. In 1930, she fractured her leg while playing at the U.S. Open at Forest Hills, New York, and played the rest of her match using a crutch, before finally losing.

Hazel Hotchkiss Wrightman is nicknamed "the Queen Mother of Tennis" because she dramatically changed women's tennis. In 1903, in San Francisco, California, she introduced volley and net play. Prior to this, the game had been played from the baseline without much movement.

In 1950, Althea Gibson became the first African American woman to play at the prestigious United States Tennis Association (USTA) at Forest Hills.

Maureen "Little Mo" Connolly of the United States was the first woman to win the Grand Slam of tennis. In 1953, she won at Wimbledon, the U.S. Open, the French Open, and the Australian Open. Only two other women have accomplished this feat: Margaret Court of Australia in 1970, and Steffi Graf of Germany in 1988.

Martina Navratilova has won more singles titles than any other tennis player in history, male or female. She briefly retired from tennis in 1994 but now competes on the doubles circuit.

In September 1999, Serena Williams won the women's singles title at the U.S. Open and became the first African American woman since Althea Gibson to win one of the Grand Slam events. The next day she teamed up with her sister, Venus, to win the U.S. Open doubles title.

In 2000, it was Venus's turn to shine as she won her first two Grand Slam singles events, Wimbledon and the U.S. Open, and added two Olympic gold medals (singles and doubles with Serena) for good measure. She continued her dominance in 2001, winning Wimbledon and the U.S. Open again. At the 2001 U.S. Open, she defeated sister Serena in the finals.

ONE WOMAN YOU'VE JUST GOT TO KNOW

American athlete **Babe Didrikson Zaharias** (1913–1956) earned more medals, broke more records, and swept more tournaments in more sports than any other athlete, male or female, in the twentieth century. She played forward with the Golden Cyclone Squad, one of the best womens' basketball teams in the country. At the 1932 Olympics, she won gold medals in javelin throwing, 80-meter hurdles, and the high jump. From 1940 to 1950, she won every available golf title. In 1945, she was named Woman Athlete of the Year in a unanimous poll of Associated Press sportswriters. In 1950, the Associated Press named her Female Athlete of the Half Century. When asked if there was anything she didn't play, Zaharias replied, "Yeah, dolls."

Science

LANDMARKS

1809 Mary Kies becomes the first American woman to receive a patent, for a method of weaving straw with silk.

1873 Ellen Swallow Richards, the first woman to be admitted to the Massachusetts Institute of Technology (MIT), earns her bachelor of science degree. She becomes the first female professional chemist in the nation.

1885 Sarah E. Goode becomes the first African American woman to receive a patent, for a bed that folds up into a cabinet. Goode, who owns a furniture store in Chicago, intends the bed to be used in apartments.

1915 The American Medical Women's Association is formed to provide a voice and a network for women in medicine. At the time, women are barred from membership in the American Medical Association.

1950 The Society of Women Engineers is founded in New Jersey. It strives to educate women about opportunities in engineering and to help female engineers reach their fullest potential.

1960 Jerrie Cobb is the first woman in the United States to undergo astronaut testing. NASA, however, cancels the women's program in 1963. It is not until 1983 that an American woman — Dr. Sally K. Ride — is sent into space.

1971 The Association for Women in Mathematics (AWM) is formed in Atlantic City, New Jersey, and the Association for Women in Science (AWIS) is organized in Chicago. They work to promote and support women's involvement in these fields.

1999 Lieutenant Colonel Eileen Collins is the first female astronaut to command a space shuttle mission. Collins was also the first woman to pilot a space shuttle, in 1995.

DID YOU KNOW?

Mystery Inventors

We'll probably never know how many female inventors there have been. That's because in colonial times and the early years of the United States, a woman could not get a patent in her own name. A patent is considered a kind of property, and until the late 1800s, laws forbade women in most states from owning property or entering into legal agreements in their own names. Instead, a woman's property was held in the name of her father or husband.

For example, many people believe that Sybilla Masters was the first American female inventor. In 1712, she developed a new corn mill, but she was denied a patent because she was a woman. Three years later, the patent was filed successfully in her husband's name.

FEMALE INGENUITY

The following is a partial list of the many ingenious devices women have invented.

INVENTION	INVENTOR	YEAR
Alphabet blocks	Adeline D. T. Whitney	1882
Apgar test, which evaluates a baby's health upon birth	Virginia Apgar	1952
Chocolate-chip cookie	Ruth Wakefield	1930
Circular saw	Tabitha Babbitt	1812
Dishwasher	Josephine Cochran	1872
Disposable diaper	Marion Donovan	1950
Electric hot water heater	Ida Forbes	1917
Elevated railway	Mary Walton	1881
Engine muffler	El Dorado Jones	1917
Fire escape	Anna Connelly	1887
Globe	Ellen Fitz	1875
Ironing board	Sarah Boone	1892
Kevlar, a steel-like fiber used in radial tires, crash helmets, and bulletproof vests	Stephanie Kwolek	1966
Life raft	Maria Beaseley	1882
Liquid Paper, a quick-drying liquid used to correct mistakes printed on paper	Bessie Nesmith	1951
Locomotive chimney	Mary Walton	1879
Medical syringe	Letitia Geer	1899
Paper bag–making machine	Margaret Knight	1871
Rolling pin	Catherine Deiner	1891
Rotary engine	Margaret Knight	1904
Scotchgard fabric protector	Patsy O. Sherman	1956
Snugli baby carrier	Ann Moore	1965
Street-cleaning machine	Florence Parpart	1900
Submarine lamp and telescope	Sarah Mather	1845
Windshield wiper	Mary Anderson	1903

She Made Beauty Big Business

One of the nation's most successful inventors was Madame C. J. Walker (1867–1919). Widowed at twenty, she supported herself and her daughter by working as a washerwoman. In the early 1900s, she invented a process for straightening the hair of African Americans. Her process, combining a unique cosmetic formula with brushes and heated combs, caught on. In 1910, she formed Madame C. J. Walker Laboratories in Indianapolis, Indiana, where she developed products and trained her beauticians, known as "Walker agents." The agents and their products became well known in black communities throughout the United States and the Caribbean. Walker's business amassed a fortune, and she became the first African American millionaire and a generous patron of many black charities.

BRAIN CHILD

A "brain child" is an original idea. Here are some girls who had great ideas that they turned into inventions.

Eight-year-old **Theresa Thompson** and her nine-year-old sister, **Mary**, were the youngest sisters to receive a U.S. patent. They invented a solar-heated tepee for a science fair project in 1960. They called the device a Wigwarm.

At age nine, **Margaret Knight** began working in a cotton mill, where she saw a steel-tipped shuttle fly out of a loom and hit a nearby worker. As a result, Margaret devised her first invention: a shuttle-restraining device. She went on to invent the machine that makes the square-bottom paper bags we still use for groceries today. That machine was patented in 1871. As an adult, she invented the rotary engine (1904).

Eleven-year-old **Jeanie Low** received a patent on March 10, 1992, for inventing the Kiddie Stool — a fold-up stool that fits under the sink so kids can unfold it, stand on it, and reach the sink on their own.

HOW TO GET A PATENT

If you think you have a great invention and want to have it patented, you must file an application with the U.S. Patent and Trademark Office. The process can be complicated. You must explain your idea clearly to a patent examiner, who will determine whether it is new and useful. You must have an illustration of the device in action. Many people hire patent attorneys, which can be very expensive, to make sure that their patent applications conform to the patent office's rules. If you get a patent, your invention will be assigned a unique number, and you alone will have the right to sell your device. For more information, contact the U.S. Patent and Trademark Office at Crystal Plaza 3, Room 2C02, Washington, DC 20231, or visit the website at www.uspto.gov.

SOME TYPES OF SCIENTISTS

An **agronomist** specializes in soil and crops.
An **astronomer** studies stars, planets, and galaxies.
A **botanist** specializes in plants.
A **cytologist** specializes in the study of cells.
An **epidemiologist** studies the spread of diseases.
An **ethologist** studies animal behavior.
A **geneticist** studies how traits are inherited.
A **geographer** studies the earth's surface.
A **geologist** specializes in the history of the earth.
A **marine biologist** studies ocean plants and animals.
A **meteorologist** studies weather and climate.
A **microbiologist** studies microscopic plants and animals.
A **paleontologist** specializes in fossils.
A **physicist** studies matter, energy, and how they are
    related.
A **seismologist** studies earthquakes.

FEMALE NOBEL PRIZE WINNERS IN SCIENCE . . .

Marie Curie
Physics, 1903, and Chemistry, 1911
Marie Curie is the only person ever to win two Nobel
Prizes. By the time she was sixteen, she had already won a
gold medal at school in Poland. In 1891, almost penniless,
she began her university education at the Sorbonne in
Paris. In 1903, her discovery of radioactivity earned her the
Nobel Prize in physics. In 1911, she won the Nobel Prize
for chemistry.

Irene Curie
Chemistry, 1935
Irene Curie was the daughter of Marie Curie. She furthered her mother's work in radioactivity and won the Nobel Prize for discovering that radioactivity could be artificially produced.

Gerty Radnitz Cori
Biochemistry, 1947
Cori was the first American woman to win a Nobel Prize in science. She studied enzymes and hormones, and her work brought researchers closer to understanding diabetes. She won the Nobel Prize for discovering the enzymes that convert glycogen into sugar and back again to glycogen.

Maria Goeppert Mayer
Physics, 1963
Mayer researched the structure of atomic nuclei. During World War II, she worked on isotope separation for the atomic bomb project.

Dorothy Crowfoot Hodgkin
Chemistry, 1964
Hodgkin discovered the molecular structures of penicillin and vitamin B_{12}. She won the Nobel Prize for determining the structure of biochemical compounds essential to combating pernicious anemia.

Rosalyn Sussman Yalow
Medicine, 1977
Yalow won the Nobel Prize for developing radioimmunoassay, a test of body tissues that uses radioactive isotopes to measure the concentrations of hormones, viruses, vitamins, enzymes, and drugs.

Barbara McClintock
Medicine, 1983
McClintock, a geneticist, studied the chromosomes in corn. Her work uncovered antibiotic-resistant bacteria and a possible cure for African sleeping sickness.

Rita Levi-Montalicini
Medicine, 1986
Levi-Montalicini is an Italian neuroembryologist known for her codiscovery in 1954 of nerve growth factor, a previously unknown protein that stimulates the growth of nerve cells and plays a role in degenerative diseases like Alzheimer's disease.

Gertrude Elion
Medicine, 1988
Elion is the only female inventor inducted into the Inventors Hall of Fame. She invented the leukemia-fighting drug 6-mercaptopurine. Her continued research led to Imuran, a derivative of 6-mercaptopurine that blocks the body's rejection of foreign tissues.

Christiane Nüsslein-Volhard
Medicine, 1995
Nüsslein-Volhard won the Nobel Prize for research that used the fruit fly to help explain birth defects in humans.

. . . AND FIVE MORE WOMEN
TO KNOW

Maria Mitchell (1818–1889), the nation's first professional female astronomer, grew up on the Massachusetts whaling island of Nantucket, among sailors who steered their ships according to the movement of the stars. She learned to chart the placement of the sun, moon, and stars as a child and charted her first eclipse when she was just twelve. She further tested her math skills at age seventeen by accurately surveying the island of Nantucket with her father. In 1847, Mitchell was working as a librarian and studying the night skies when she discovered a comet, which became known as "Miss Mitchell's Comet" and earned her a gold medal from the king of Denmark. The following year Mitchell became the first female member of the American Academy of Arts and Sciences. From 1865 to 1888, she taught science to hundreds of young women as professor of astronomy at Vassar College in New York.

Mary Leakey (1913–1996) began participating in archaeological digs in England at age seventeen. She never attended university. In 1935, she and her husband, Louis Leakey, moved to Kenya, where they began a thirty-year collaboration to uncover evidence of early hominids. In 1948, Mary found a perfectly preserved skull of *Proconsul africanus,* an apelike ancestor of humans. In 1959, she discovered a hominid fossil in Olduvai Gorge, Tanzania, that was thought to be 1.7 million years old. Her most important discovery, however, took place in 1978, when her team uncovered the footprints of two hominids in Laetoli, Tanzania. The footprints, thought to be 3.5 million years old, indicated that humans began walking upright much earlier than scientists had previously thought.

Dr. Jane Goodall (b. 1934) is the world's leading authority on chimpanzees. At the age of twenty-three, determined to learn about African wildlife, she left her native England for East Africa, where she hunted for fossils with Louis and Mary Leakey. Three years later, accompanied only by her mother, she began to study chimpanzees at Gombe National Park in Tanzania. The chimpanzees at Gombe gradually accepted Goodall, and, with patient observations, she learned much about them. Before Goodall's work, no one knew just how similarly chimpanzees and humans behave — for example, that chimpanzees make and use tools, or that they eat meat. More than forty years later, Goodall continues her Gombe work, which is now the longest field study of any wild animal species.

Rear admiral **Dr. Grace Murray Hopper** (1906–1992) was one of the earliest computer programmers and a leader in the field of software development concepts. She joined the United States Naval Reserve in 1943 and was assigned to the Bureau of Ordnance Computation Project at Harvard University, where she worked at Harvard's Cruft Laboratories on the Mark series of computers. In 1949, she

DID YOU KNOW?

"Female Doctor" Was Once a Radical Idea

Although women have always practiced medicine — as health experts in their families, as healers in their communities — their opportunities to become professional doctors were very limited for many years. It was once common for medical schools to refuse to admit women. Harvard Medical School did not accept women until 1945, and only then because World War II had reduced the number of men enrolled as students. That was almost one hundred years after women first had applied to the school!

Men are still in the majority in the medical professions. In 1998, women made up just 23 percent of practicing doctors. But times are changing. It is estimated that by the year 2010, at least half of the medical school students in the nation will be women.

joined the Eckert-Mauchly Computer Corporation, subsequently the Sperry Corporation, as a senior mathematician, while still a consultant and lecturer for the United States Naval Reserve. After retiring from the Navy in 1986, she became a senior consultant to Digital Equipment Corporation and continued working well into her eighties.

When **Dr. Mae Jemison** (b. 1956) completed her NASA training in 1988, she became the first African American female astronaut. Four years later she became the first African American woman in space when she left Earth on a joint U.S.–Japanese space mission. Before training as an astronaut, Jemison earned a bachelor's degree in chemical engineering and a doctorate in medicine. As a doctor, she has treated people on several continents, spending two and a half years with the Peace Corps in West Africa, where she supervised medical staff and wrote health care manuals. She has also worked with the Centers for Disease Control and Prevention to develop vaccines. Jemison is a college professor and the founder and president of two technology companies.

DISEASE FIGHTERS

Women in medicine have helped improve human health in many ways. Here are just a few of the many diseases female scientists have helped to fight.

AIDS

In 1988, a team of researchers led by the chemist Jane Rideout won a patent for AZT, a drug used to fight acquired immune deficiency syndrome (AIDS).

☞ SHE SAID IT ☜

"You know, there are people who believe that science will answer everything, that every little mystery will be explained. I would be very, very surprised — not in my lifetime, I don't think in anybody's lifetime. The more science teaches us about this universe, the more amazing it is."

— Jane Goodall, British ethologist

"Those who dwell, as scientists or laymen, among the beauties and mysteries of the earth are never alone or weary of life."

— Rachel Carson, American marine biologist

"We have a hunger of the mind which asks for knowledge of all around us, and the more we gain, the more is our desire; the more we see, the more we are capable of seeing."

— Maria Mitchell, American astronomer

"No two plants are exactly alike. They're all different, and as a consequence, you have to know that difference. I start with the seedling, and I don't want to leave it. I don't feel I really know the story if I don't watch the plant all the way along. So I know every plant in the field. I know them intimately, and I find it a great pleasure to know them."

— Barbara McClintock, American geneticist and Nobel laureate

"Nothing in life is to be feared. It is only to be understood."

— Marie Curie, French-Polish chemist and Nobel laureate

Brucellosis

Alice Evans (1881–1975) discovered the cause of the deadly disease brucellosis, also known as undulant fever. The disease was caused by contaminated milk.

Cystic Fibrosis

Dorothy Hansine Anderson (1901–1963) was the first person to identify cystic fibrosis. She devised an easy method of diagnosing the disease in its early stages.

Hodgkin's Disease

Dorothy Mendenhall (1874–1964) identified the cell used to diagnose Hodgkin's disease.

Meningitis

Hattie Elizabeth Alexander (1901–1968) developed the meningitis serum. Before her discovery, this disease was 100 percent fatal in infants.

Mononucleosis

Karen Elizabeth Waillard-Gallo invented and patented a method for early detection of infectious mononucleosis.

Polio

Dorothy Horstmann (b. 1911) identified the polio virus in its early stages. Her discovery was an important factor in developing a vaccine.

Rheumatic Fever

Rebecca Lancefield (1895–1981) is credited with first categorizing the organism responsible for rheumatic fever.

Arts and
Entertainment

Art and Architecture

LANDMARKS

1707 Henrietta Johnston begins to work as a portrait artist in Charles Town (now Charleston), South Carolina, making her the first known professional female artist in America.

1848 The Pennsylvania Academy of Fine Arts becomes the first art school in the United States to accept women as students.

1868 For the first time, the "Ladies Life" class at the Pennsylvania Academy of Fine Arts allows women to draw from a live nude model, as male artists have done for centuries.

1890 Louise Blanchard Bethune is the first woman to become a full member of the American Institute of Architects.

1933–1943 Photographer Dorothea Lange and painter Alice Neel are some of the many artists who receive federal support for their art through New Deal programs.

1987 The National Museum of Women in the Arts opens in Washington, D.C.

THROUGH HER OWN EYES

Every artist has a unique vision. Here are some long-popular subjects that took on new looks through the eyes of female artists.

Flowers Giant flowers painted by American artist Georgia O'Keeffe (1887–1986) feature extreme close-ups that make each flower look like a world unto itself.

Self-Portraits The self-portraits of Mexican painter Frida Kahlo (1911–1954) include surreal touches, such as body parts combined with parts of plants or animals.

Landscape Folk artist Grandma Moses (1860–1961) crowded her amazingly detailed landscapes with every activity and creature they could possibly hold.

Water Some paintings by American artist Helen Frankenthaler (b. 1928) depict swells of water that seem to bloom and flow across her giant canvases.

Nudes American painter Alice Neel (1900–1984) brought a new realism to the classic study of the nude, depicting the effects of light, gravity, and age on the human form.

Mother and Baby American painter Mary Cassatt (1844–1926) found beauty in mothers and children as they engaged in everyday activities, such as a gentle bath.

Still Life The glassware and other table items in the work of American painter Janet Fish (b. 1938) are so rich with light and shadow that they seem to glow from within.

FIVE PHOTOGRAPHERS TO KNOW

Julia Margaret Cameron (1815–1879) dressed her family, friends, and visitors in interesting costumes and photographed them in dramatic scenes. Some of her imaginative, romantically lit images resemble Renaissance paintings.

Dorothea Lange (1895–1965) is renowned for her haunting photographs of migrant workers, farmers, and other Americans who suffered through the Great Depression. Her pained *Migrant Mother* (1936) became the image that defined the era.

Berenice Abbott (1898–1991) photographed artists and writers in Paris in the 1920s and brought her craft back to her native New York in the 1930s. Her images of New York street life, from grocery stores to fish markets to subways, are among her best-loved work.

Diane Arbus (1923–1971) spent years as a successful fashion photographer before turning her eye to people on the fringes of society. Circus performers and patients at mental institutions are among the subjects of her bold and brightly lit photographs.

Annie Leibovitz (b. 1949) is a photographer of celebrities who has become a celebrity herself. At age twenty-three, she became chief photographer for *Rolling Stone* magazine, where she worked from 1973 to 1983. She remains the best-known photographer of her generation.

SOME FAMOUS STRUCTURES
BY FEMALE ARCHITECTS

The Vietnam Veterans Memorial (1980–1982) in Washington, D.C., by Maya Lin
   Two long, low, black granite walls set into the ground are inscribed with the names of the thousands lost in the war.

Hearst Castle (1919–1948) in San Simeon, California, by Julia Morgan

Morgan worked for decades on this fantastical complex of Spanish-style buildings, which includes numerous houses, monumental stairways, and a Roman-style pool.

The Rosenthal Center for Contemporary Art (2000–2003) in Cincinnati, Ohio, by Zaha Hadid
The boxy galleries of this museum, made up of translucent interior and exterior walls, seem to float in a zigzag pattern above the street.

The Musée d'Orsay (opened 1987) in Paris, renovated by Gae Aulenti
This old Paris train station was converted into a museum for modern art, with galleries branching off from the hall where train tracks once ran.

Music and Dance

LANDMARKS

1792 Suzanne Vaillande appears in *The Bird Catcher,* the first ballet presented in the United States, in New York. She was also probably the first woman to work as a choreographer and set designer in the United States.

1880s–1900s All-female orchestras, such as the Boston Fadette Lady Orchestra, become popular acts in American cities.

1897 H. H. A. Beach's *Gaelic Symphony* is the first symphony by a woman performed in the United States, and possibly the world.

1914 Mary Davenport-Engberg is the first woman to conduct a symphony orchestra, in Bellingham, Washington.

1972 Women dominate the 1971 Grammy Awards, taking all four top categories. Carole King wins record, album, and song of the year, while Carly Simon takes the best new artist award.

1973 Singer Patsy Cline becomes the first woman inducted into the Country Music Hall of Fame.

1974 Patti Smith releases what is considered to be the first punk rock single, "Hey Joe."

1980s and beyond MTV helps launch the careers of a number of superstars, including Cyndi Lauper, Madonna, and Janet Jackson.

1983 Ellen Taafe Zwilich becomes the first woman awarded the Pulitzer Prize for music, for *Three Movements for Orchestra*.

1987 Singer Aretha Franklin becomes the first woman inducted into the Rock and Roll Hall of Fame.

1998 The Lilith Fair, an all-female music tour, becomes one of the most successful musical events of the year.

SING IT, SISTER

Women penned these famous song lyrics.

"Twinkle, twinkle, little star
How I wonder what you are"
— "The Star" (1806), by Ann Taylor

"Swing low, sweet chariot
Coming for to carry me home"
— "Swing Low, Sweet Chariot" (1847),
by Sarah Hannah Sheppard

"Mine eyes have seen the glory of the coming
of the Lord"
— "The Battle Hymn of the Republic"
(1862), by Julia Ward Howe

"O beautiful for spacious skies
For amber waves of grain"
— "America the Beautiful" (1893),
by Katherine Lee Bates

"Mama may have, Papa may have
But God bless the child that's got his own"
> — "God Bless the Child" (1941),
> by Billie Holiday

"When my soul was in the lost-and-found
You came along to claim it"
> — "Natural Woman" (1967), by Carole King

"I've looked at life from both sides now,
From win and lose, and still somehow
It's life's illusions I recall.
I really don't know life at all"
> — "Both Sides Now" (1969), by Joni Mitchell

"I'm proud to be a coal miner's daughter
I remember well the well where I drew water"
> — "Coal Miner's Daughter"
> (1970), by Loretta Lynn

"Workin' 9 to 5
What a way to make a livin'"
> — "9 to 5" (1980), by Dolly Parton

SINGING VOICES

A singing voice is classified by how high or low a performer naturally sings. Here are the classifications for female voices.

Classification	Range	Example
contralto (or "alto")	low	Marian Anderson
mezzo-soprano	middle	Cecelia Bartoli
soprano	high	Jessye Norman

FIVE SINGERS TO KNOW

Legendary contralto **Marian Anderson** (1897–1993) was the first African American member of New York's Metropolitan Opera. She also made history in 1939 when, after being banned from performing at Constitution Hall because she was black, she gave a concert on the steps of the Lincoln Memorial that was attended by seventy-five thousand people.

Jazz singer **Ella Fitzgerald** (1918–1996) is considered by many to be the greatest jazz vocalist of all time and is credited with influencing a generation of singers. Her hits include "A-Tisket-A-Tasket" (1928) and "Lady Be Good" (1947). In a career that spanned six decades, Fitzgerald recorded more than 250 albums with jazz's elite players.

Soprano **Joan Baez** (b. 1941) brought folk music into the mainstream with her best-selling albums and beautiful renditions of traditional ballads and spirituals. Her first album, self-titled *Joan Baez* (1960), collected thirteen such songs. Baez was one of the first politically outspoken performers, an advocate for civil rights and the peace movement.

Soul singer **Aretha Franklin** (b. 1942), "the Queen of Soul," is also a grand diva of pop music. Her unmistakable style fuses gospel, jazz, rock, and rhythm and blues, especially in such late-1960s hits as "Respect" (1967), "Chain of Fools" (1968), and "I Say a Little Prayer" (1969). Franklin has sold millions of albums and won fifteen Grammy Awards.

Soprano **Barbra Streisand** (b. 1942) began performing as a teenager and became a star via the stage (1964) and screen (1968) versions of *Funny Girl,* a musical based on the life of comedienne Fanny Brice. The multitalented singer, actress, composer, and director has had a staggering number of best-selling albums and has won several Oscars and Grammys.

GREAT BALLET CHARACTERS

Because many ballets take their stories from folklore, they can feel like fairy tales told through dance. Here are the heroines of some magical favorites.

Clara in *The Nutcracker*
A nutcracker doll Clara receives for Christmas leads the way into an enchanted world.

Coppelia in *Coppelia*
A doll is so exquisitely beautiful and lifelike that she wins the heart of a village boy.

Giselle in *Giselle*
A young girl spends her days dancing and gathering flowers until love brings tragedy.

Odette in *Swan Lake*
Under a magic spell, the swan Odette becomes a human each night — and falls in love.

Ondine in *Ondine*
A curious young water nymph pays an unforgettable visit to the land of mortals.

Princess Aurora in *The Sleeping Beauty*
One fairy's curse and another fairy's gift mark the destiny of a young princess.

DID YOU KNOW?

Modern Women

Female choreographers such as Isadora Duncan, Ruth St. Denis, and Martha Graham were leaders in the development of modern dance, which was created in the early twentieth century. Modern dance was a rejection of the traditions of ballet. It went against what some saw as ballet's rigid steps, limited emotional expression, and dainty sense of beauty — especially for women. The often informal look of modern dance, and the strength and originality displayed by modern dancers, yielded new perceptions of female dancers as well as new horizons for female choreographers.

SOME DANCE CRAZES

Boogie-woogie

Cha-cha

Fox-trot

Funky chicken

Jitterbug

Lindy hop

Rumba

Twist

Cakewalk

Fandango

Frug

Hustle

Jive

Macarena

Shimmy

MOVIES

LANDMARKS

1896 Alice Guy Blaché, the first American female film director, shoots the first of her more than three hundred films, a short feature called *La Fee aux Choux* (The Cabbage Fairy).

1910s–1920s Theda Bara, "the Vamp," and Mary Pickford, "America's Sweetheart," are some of the first stars of the silent screen. Pickford also founds her own studio and is a longtime producer.

1927 The silence of the screen is broken, as technological improvements allow actors to speak. Actresses Greta Garbo and Marlene Dietrich are among the first stars in the new "talkies."

1929 The first Academy Awards ceremony takes place in Hollywood, California. Janet Gaynor wins the best actress award for her role in *Seventh Heaven*.

1935 *Becky Sharp* is the first Technicolor film, with Miriam Hopkins in the title role.

2002 Halle Berry becomes the first African American to win an Academy Award for best actress, for her role in *Monster's Ball*.

Classic Quotes in the Movies

"Life is a banquet, and most poor suckers are starving to death."

> — Auntie Mame (played by Rosalind Russell),
> in *Auntie Mame* (1958)

"Don't be such a beetle! I could never love anyone as I love my sisters!"

> — Jo March (played by Winona Ryder),
> in *Little Women* (1994)

"Home! And this is my room, and you're all here. And I'm not gonna leave here ever, ever again, because I love you all, and — oh, Auntie Em — there's no place like home!"

> — Dorothy (played by Judy Garland),
> in *The Wizard of Oz* (1939)

"Marriage, fun? Fiddle-dee-dee! Fun for men, you mean."

> — Scarlett O'Hara (played by Vivien Leigh),
> in *Gone With the Wind* (1939)

FIVE CHILD ACTRESSES TO KNOW

Shirley Temple (b. 1928) began dancing almost as soon as she could walk. She was only four in her first film, *The Red-Haired Alibi* (1932). At the age of six, she was awarded a special Oscar for her performance in the movie *Bright*

SHE SAID IT

Classic Quotes in the Movies, continued

"But, Captain, whistles are for animals, not for children, and definitely not for me!"

> — Maria (played by Julie Andrews),
> in *The Sound of Music* (1965)

"Real diamonds! They must be worth their weight in gold!"

> — Sugar (played by Marilyn Monroe),
> in *Some Like It Hot* (1959)

"Oh, we're going to talk about me again, are we? Goody!"

> — Tracy (played by Katharine Hepburn),
> in *The Philadelphia Story* (1940)

Eyes. From 1935 to 1938, she was the most popular U.S. box-office attraction. She stopped making movies in 1949, but in the 1960s, she began a political career — as a diplomat and the first woman appointed chief of protocol in the U.S. State Department.

Judy Garland (1922–1969) was seven when she began singing in vaudeville shows with her two sisters. She signed a movie contract at age thirteen, and soon afterward starred as Dorothy in *The Wizard of Oz* (1939), her most enduring film. She continued to sing in movie musicals through the 1940s. When she died in 1969 at age forty-seven, she had become a musical legend. More than twenty thousand people attended her funeral.

Elizabeth Taylor (b. 1932) was born in England and began her acting career at age ten with a small part in the movie *There's One Born Every Minute* (1942). She became a major star at age twelve in the classic horse movie *National Velvet* (1944), and at seventeen played the role of Amy in the first film version of *Little Women* (1949). Taylor is known for her extraordinary beauty, long career, and nine marriages.

Jodie Foster (b. 1962) began her prolific career at age two, appearing in commercials and television programs. Her first movie, the Disney film *Napoleon and Samantha*, was released in 1972. She gained fame (and her first Oscar nomination) for a controversial role in *Taxi Driver* (1976) when she was just thirteen. She continued her acting career while earning a degree (with honors) from Yale. In addition to starring in films such as *Silence of the Lambs* (1991), Foster has directed several films, including *Little Man Tate* (1991).

Drew Barrymore (b. 1975) was born into the legendary Barrymore family of actors. She made her first commercial at nine months and became a star in *E.T.: The Extra-Terrestrial* (1982) at age seven. She continues to work prolifically, turning out several performances a year in films ranging from musicals (*Everyone Says I Love You*, 1996) to romances (*Ever After*, 1998) to action films (*Charlie's Angels*, 2000).

SCREEN LEGENDS

In 1999, the American Film Institute named its greatest screen legends of all time — actors and actresses with "a significant screen presence in feature-length films" who made their debuts before 1950. Here are the top ten actresses on their list.

1. Katharine Hepburn
2. Bette Davis
3. Audrey Hepburn
4. Ingrid Bergman
5. Greta Garbo
6. Marilyn Monroe
7. Elizabeth Taylor
8. Judy Garland
9. Marlene Dietrich
10. Joan Crawford

DID YOU KNOW?

There Was Entertainment before Movies

Before movies, television, and even radio, one of the most popular forms of entertainment was the vaudeville theater. Vaudeville combined song, dance, magic acts, and funny skits. Between about 1880 and 1930, there were hundreds of vaudeville theaters across the nation. Bawdy blond actress Mae West and singing comedienne Fanny Brice were two of the women who first became famous through vaudeville.

"I found I could say things with color and shapes that I couldn't say in any other way — things I had no words for."

— Georgia O'Keeffe, American painter

"I really believe there are things nobody would see if I didn't photograph them."

— Diane Arbus, American photographer

"The truest expression of a people is in its dance and music."

— Agnes De Mille, American choreographer

"Being a singer is a natural gift. It means I'm using to the highest degree possible the gift that God gave me to use. I'm happy with that."

— Aretha Franklin, American singer

"You'd think [acting] is something one would grow out of. But you grow into it. The more you do, the more you realize how painfully easy it is to be lousy and how very difficult to be good."

— Glenda Jackson, British actress

Careers

LANDMARKS

1824 The first American public high school for girls opens in Worcester, Massachusetts.

1830s Mills in industrial towns such as Lowell, Massachusetts, are staffed almost entirely by young women. These "mill girls" have a kind of independence their mothers could not have imagined. They earn their own money and live together in boardinghouses. They also take part in strikes and other actions of organized labor.

1837 Oberlin College, in Ohio, becomes the first college to admit female students. In addition to studying, the women have to do laundry and cook meals for the male students.

1837 The Mount Holyoke Female Seminary, in Massachusetts, is created to provide higher education to women. Most of its early graduates become teachers.

1860s–1890s A number of colleges for women are opened: Vassar (1865) in New York State; Wellesley (1875), Smith (1875), and Radcliffe (1893) in Massachusetts; Bryn Mawr (1885) in Pennsylvania; and Barnard (1889) in New York City. With Mount Holyoke, which becomes a college in 1893, they are known as the Seven Sisters.

1881 Spelman College, the first college for black women in the United States, is founded in Atlanta, Georgia.

1942–1945 Thousands of women fill the new jobs created by World War II. "Rosie the Riveter" becomes a symbol of the glamorous and patriotic female factory worker. But when the war ends, the Rosies are sent home — or back to the low-paying agricultural or domestic work they had done before the war.

1963 The Equal Pay Act makes it illegal for companies to pay different rates to women and men who do the same work.

1963 The Presidential Commission on the Status of Women, appointed by President Kennedy and chaired by Eleanor Roosevelt, releases a report that details inequalities faced by women. Kennedy follows with a presidential order demanding that the civil service make hiring decisions "solely on the basis of ability" and "without regard to sex."

1964 Title VII of the Civil Rights Act of 1964 makes it illegal for employers to discriminate on the basis of race or sex. Never before had it been illegal for a company to

DID YOU KNOW?

Separate and Unequal

Until the early 1960s, newspapers published separate job listings for men and women. Jobs were categorized according to sex, with the higher-level jobs listed almost exclusively under "Help Wanted — Male." In some cases employers ran identical jobs in the male and female listings — but with separate pay scales. Separate, of course, meant unequal: between 1950 and 1960, women with full-time jobs earned on average between 59¢ and 64¢ for every dollar their male counterparts earned in the same jobs.

refuse to hire or promote a woman just because of her sex! The Equal Employment Opportunity Commission (EEOC) is created to enforce the new law. Nevertheless, women and groups such as the National Organization for Women (NOW), formed in 1966, must force the EEOC to correct specific instances of sex discrimination.

1974 Two groundbreaking organizations are formed to improve conditions in secretarial work, a mostly female field: 9 to 5, in Boston, and Women Employed, in Chicago. As in traditional labor unions, members work for better wages and job security. But they also use the media to reveal the disrespect faced by many female office workers. For instance, when women are fired for refusing to make coffee or run personal errands for their bosses, the groups organize pickets to publicize the insulting treatment.

1974 Women from such labor unions as the United Auto Workers come together to found the Coalition of Labor Union Women (CLUW). CLUW helps give voice to women and the problems they face within their various unions. It also helps women to achieve recognition as a significant force within the labor movement as a whole.

DOCTOR, LAWYER, INDIAN CHIEF

You can be almost anything you want to be. Here is a sampling of just some of the careers you might consider, along with women who have succeeded in them.

Ambassador Eugenie Anderson (1910–1997), the first female U.S. ambassador and the first woman to sign a treaty on behalf of the United States, served as ambassador to Denmark from 1949 to 1953.

Architect When she was twenty-one, Maya Lin (b. 1960) won a national competition to design and build the now famous Vietnam Veterans Memorial in Washington, D.C.

Astronaut Doctor, astronaut, and Peace Corps veteran Mae Jemison (b. 1956) became the first African American woman to enter space when she served on the crew of the space shuttle *Endeavor* in September 1992.

Astronomer Annie Jump Cannon (1863–1941), one of the greatest astronomers of the twentieth century, discovered hundreds of stars and classified about five hundred thousand.

Aviator Bessie Coleman (1893–1926) was the first African American woman to receive a pilot's license and the first woman to receive an international pilot's license.

Ballerina American Indian Maria Tallchief (b. 1925) was a prima ballerina at the New York City Ballet for many years, as well as a founder of the Chicago City Ballet.

Bishop In 1989, Barbara Harris (b. 1930) was consecrated a bishop of the U.S. Episcopal Church, the first woman and one of the first African Americans to hold that position.

Botanist Ynes Mexia (1870–1938) collected plant specimens, many of them never before identified, in remote areas from Alaska to the Amazon to the Andes Mountains.

Chef American-born Julia Child (b. 1912) popularized French cooking in the United States with her television show, *The French Chef,* in the 1960s.

Chief Wilma Mankiller (b. 1945), a longtime activist for Native American rights, served as chief of the Cherokee Nation from 1985 to 1995. She was the first woman in modern history to lead a major Native American tribe.

Conductor Eve Queler (b. 1936) has conducted numerous orchestras and more than sixty operas worldwide, becoming one of the few women to be addressed as "maestro."

Cowgirl Johanna July (1850?–1930?), born to a family of Seminole Indians and former slaves, was known throughout Texas for her ability to tame wild horses.

Director French-American Alice Guy Blaché (1875–1968) was the first American female film director and one of the first directors to work with color and sound.

Diva Soul singer Aretha Franklin (b. 1942) has been a legend for more than forty years. The Michigan legislature once declared her voice one of the state's greatest natural resources.

Diver American diver Pat McCormick (b. 1930) won women's platform and springboard gold medals in both the 1952 and 1956 Olympics.

Doctor Mary Edwards Walker (1832–1919) was a commissioned assistant surgeon for the Union army during the Civil War and is the only woman ever to be awarded the Medal of Honor, the nation's highest military award.

Environmentalist Rachel Carson (1907–1964) helped launch the environmental protection movement with her most famous book, *Silent Spring,* which changed how many Americans thought about pesticides.

General Brigadier General Wilma Vaught (b. 1930), one of the most decorated women in U.S. military history, was the first female general in the Air Force. She was also a planner for the Women in Military Service for America Memorial in Washington, D.C.

Heptathlete Jackie Joyner-Kersee (b. 1962) may be the all-time greatest competitor in the heptathlon — a sport comprising seven different track-and-field events. She has won three gold, one silver, and one bronze Olympic medals.

Ichthyologist Marine biologist and skin diver Eugenie Clark (b. 1922), "the Shark Lady," has shared her lifelong love of fish in three books and many television specials.

Illustrator To create the eerie goblins and mysterious figures depicted in her books, Molly Bang (b. 1943) draws from the folktales she gathered in her worldwide travels.

Interior Designer The influential Elsie de Wolfe (1865–1950), generally considered the first American interior designer, popularized a fresh, airy look that included comfortable sofas, gilded mirrors, and light colors.

Jockey In 1970, Diane Crump (b. 1949) became the first woman to ride in the Kentucky Derby, leading the way for other female professional riders.

Journalist Anna Louise Strong (1885–1970) covered revolutions in China and Russia and traveled all over Asia, including areas, such as Tibet and Laos, that few westerners had seen.

Judge Ruth Bader Ginsburg (b. 1933), who was appointed to the U.S. Supreme Court in 1993, has advanced women's rights during her impressive career by successfully arguing a number of sex discrimination cases.

Labor Leader Dolores Huerta (b. 1930) is a founder of the United Farm Workers, a labor union that helped farm workers organize and bargain for better wages and working conditions.

Lawyer Arabella Mansfield (1846–1911), the nation's first female lawyer, passed the Iowa bar exam in 1869 despite the fact that she never attended law school.

Meteorologist During World War II, Joanna Simpson (b. 1923) used her weather expertise to help plan battles. Later, as a chief scientist for NASA, she conducted research that made modern air flight safer.

Military Leader Mary Hallaren (b. 1907) led the first Women's Army Corps (WAC) battalion, a noncombatant force, in World War II and directed the organization after the war ended.

Pacifist Writer and peace activist Dorothy Day (1897–1980) protested wars and weaponry and helped found the *Catholic Worker,* an influential pacifist newspaper.

Paleontologist Sue Hendrickson (b. 1949) made headlines for finding the largest, best-preserved *Tyrannosaurus rex* yet discovered. The skeleton, nicknamed "Tyrannosaurus Sue," was mounted at Chicago's Field Museum in 2000.

Philosopher The French philosopher Simone de Beauvoir (1908–1986) became famous in 1949 when she published her book *The Second Sex,* which traced the oppression of women throughout history using her theories of psychology and myth.

Photographer Margaret Bourke-White (1906–1971), a photographer whose World War II photographs for *Life* magazine became world famous, was the creator of the photo essay, a form in which a series of photos tells a story.

Poet During her varied career, Rita Dove (b. 1952) has published books of poetry about her own family life and travels, and she has served as poet laureate of the United States.

Publisher Katharine Graham (1917–2001) became the publisher of the *Washington Post,* one of the most powerful and influential newspapers in the United States, in 1969.

Representative in Congress Patsy Takemoto Mink (1927–2002), a Democrat from Hawaii, became the first Asian American congresswoman when she was elected to the U.S. House of Representatives in 1965. She served for a total of twenty-four years.

Scientist The most famous female scientist, Marie Curie (1867–1934), is the only person to have won two Nobel Prizes — one for physics (1903) and one for chemistry (1911).

Sculptor Louise Nevelson (1900–1988) created huge, intriguing sculptures from found objects, including rough wood, broken mirrors, electric lights, and metal factory parts.

Senator When Nancy Kassebaum (b. 1932) was elected to the Senate from Kansas in 1978, she was the only woman there. She served for eighteen years.

Sled-Dog Racer Susan Butcher (b. 1956) is a four-time

winner of the Iditarod dogsled race (in 1986, 1987, 1988, and 1990).

Teacher Mary McLeod Bethune (1875–1955) devoted her life to teaching and founded the school that became Florida's historically black Bethune-Cookman College.

☞ SHE SAID IT ☜

"Just don't give up trying to do what you really want to do. Where there is love and inspiration, I don't think you can go wrong."

— Ella Fitzgerald, American singer

"Never work just for money or for power. They won't save your soul or help you sleep at night."

— Marian Wright Edelman, American founder and president of the Children's Defense Fund

"It had long since come to my attention that people of accomplishment rarely sat back and let things happen to them. They went out and happened to things."

— Elinor Smith, early American pilot and journalist

"To love what you do and feel that it matters—how could anything be more fun?"

— Katharine Graham, former publisher of the *Washington Post*

"One never notices what has been done; one can only see what remains to be done."

— Marie Curie, French-Polish chemist

She lectured widely on the necessity of education and served as adviser to three presidents.

Undersea Explorer Sylvia Earle (b. 1936), an environmental activist and marine botanist, has explored depths of more than one thousand feet and once lived in an underwater research center for two weeks.

Veterinarian The first two women to graduate from veterinary school, Elinor McGrath and Florence Kimball, did so in 1910. They were also among the first veterinarians to specialize in pet care.

Zoologist A researcher who was fascinated by invertebrates (animals, such as jellyfish, that do not have spines), Libby Hyman (1888–1969) wrote the landmark six-volume *Encyclopedia of Invertebrates.*

❈

SHUT OUT OF SCHOOL

In colonial days, few girls were taught to read. It has been estimated that less than half of them could even write their own names. Girls were not commonly educated until the end of the 1800s, when free public schooling became widely available.

But even then, there were fears that schooling could be bad for girls and women, especially at the college level. Some doctors worried that education was unnatural for women and could make them hysterical and unable to bear children. There was also doubt about whether the female brain could understand subjects like Latin and math. Meanwhile, without training, women had few opportunities to choose interesting and profitable careers.

WORK IN THE
VERY OLDEN DAYS

If you are wondering what career choices women had in the past, take a look at these.

Women's Work during the Stone Age

Making clothes from animal skins
Making shelters
Making clay pots and vessels
Gathering roots, berries, and other plant foods
Cooking
Caring for children

Women's Roles in Ancient Egypt

Midwife

Merchant

Weaver

Potter

Temple dancer

Musician

Artist

Poet

Priestess

North American Women's Work in 1492

Weaving baskets
Gathering food
Building homes
Sowing and harvesting corn and potatoes
Catching and cooking fish and shellfish
Cultivating berries and herbs for use as medicines
Fashioning fur and leather into clothing
Healing with traditional medicines

DID YOU KNOW?

It Really Does Pay to Learn

Education not only makes your life richer; it can make your bank account richer, too. As you'll see, there is a link between how much education you have had and how much you are likely to earn.

American women who have . . .	earn an average annual salary of . . .
Not graduated from high school	$12,145
A high school diploma	$18,092
An associate's degree	$25,079
A bachelor's degree	$32,546
A master's degree	$42,378
A professional degree	$59,792
A doctoral degree	$61,136

Women Who Rule

LANDMARKS

1872 Victoria Claflin Woodhull becomes the first female presidential candidate in the United States when she is nominated by the National Radical Reformers.

1916 Jeannette Rankin of Montana is the first woman to be elected to the U.S. House of Representatives.

1922 Rebecca Felton of Georgia is appointed to the U.S. Senate to fill a temporary vacancy. The first female senator (and the oldest, at eighty-seven), she serves for only two days — and gives a speech to her fellow senators, in which she predicts that they will see other women in the Senate.

1925 Nellie Tayloe Ross becomes the first woman to serve as governor of a state, in Wyoming. She is chosen to succeed her deceased husband, William Bradford Ross.

1932 Hattie Caraway of Arkansas becomes the first woman elected to the U.S. Senate.

1933 Frances Perkins is appointed U.S. secretary of labor by President Franklin D. Roosevelt, making her the first female member of a presidential cabinet.

1964 Margaret Chase Smith of Maine becomes the first woman nominated for president of the United States by a major political party, at the Republican National Convention in San Francisco.

1965 Patsy Takemoto Mink of Hawaii becomes the first Asian American woman elected to Congress. She serves in the U.S. House of Representatives for twenty-four years.

1969 Shirley Chisholm of New York becomes the first African American Congresswoman. She serves in the U.S. House of Representatives for fourteen years.

1981 Sandra Day O'Connor is appointed by President Ronald Reagan to the Supreme Court, making her its first female justice.

1985 Wilma Mankiller becomes the first female chief of the Cherokee Nation of Oklahoma.

1989 Ileana Ros-Lehtinen of Florida becomes the first Hispanic woman elected to Congress. She serves in the U.S. House of Representatives.

1992 Carol Moseley-Braun of Illinois becomes the first African American woman elected to the U.S. Senate.

1993 Janet Reno becomes the first female U.S. attorney general.

1997 Madeleine Albright is sworn in as U.S. secretary of state. She is the first woman in this position as well as the highest-ranking woman in the United States government.

2001 Condoleezza Rice becomes the first woman to serve as national security adviser.

WOMEN IN U.S. GOVERNMENT

There are a record number of women serving in the 108th (2003–2005) Congress: fifty-nine in the House and thirteen in the Senate. They represent about 14 percent of Congress as a whole. This is an improvement: over our

DID YOU KNOW?

Women Have Led Many Nations

Although the United States has not yet elected a female president, women have served as presidents and prime ministers in countries on every continent (except, of course, Antarctica!). Bangladesh, Canada, Indonesia, Ireland, New Zealand, and Turkey are among the countries that have recently had women as leaders.

nation's history, women have made up only about 2 percent of the more than two thousand congresspersons.

A total of forty-five women have been appointed or elected to Congress to fill seats left open after the deaths of their husbands: thirty-seven in the House and eight in the Senate.

Twenty-nine women have held cabinet or cabinet-level positions. Almost all of these have served after 1980. Some departments — such as the Department of the Treasury and the Department of Defense — have not yet been headed by women.

A record number of women — four — were elected state governors in 2002. This brought the total of women who have served as state governors in all of U.S. history up to twenty-three, compared with more than two thousand men.

All U.S. presidents and vice presidents have been men. But that could change in your lifetime! In a 1999 Gallup poll, 92 percent of respondents said they would vote for a qualified female candidate. In 1937, only 33 percent said they would.

FEMALE POLITICAL LEADERS — HISTORICAL AND CURRENT

Throughout world history, many women have ruled their countries, as queens, presidents, and prime ministers. In some cases, the queens who are alive and in power today do not actually rule their countries; they are symbolic monarchs only. This is a list of female rulers, their countries, and the dates they ruled.

COUNTRY	NAME	REIGN
Angola	Queen Nzingha	1582–1663
Argentina	President Isabel Perón	1974–1976
Armenia, Lesser	Queen Zabel	1219–1226
Bangladesh	Prime Minister Khaleda Zia	1991–1996, 2001–
	Prime Minister Sheikh Hasina Wajed	1996–2001
Barbados	Governor-General Dame Nita Barrow	1990–1995
Belize	Governor-General Dame Minita Gordon	1981–1993
Bermuda	Premier Pamela Gordon	1997–1998
	Premier Jennifer Smith	1998–
Bolivia	Prime Minister Lidia Gueiler	1979–1980
Brazil	Queen Maria I	1815–1816
	Empress Isabel (regent)	1871–1872, 1876–1877, 1887–1888
Burundi	Prime Minister Sylvie Kinigi	1993–1994
Byzantium (Roman Empire)	Empress Theodora	1055–1056
Cambodia	Queen Ang Mey	1835–1840, 1844–1845
	Queen Kossamak (joint ruler)	1955–1960
Canada	Governor-General Jeanne Sauvé	1984–1990
	Prime Minister Kim Campbell	1993 (4 months)

COUNTRY	NAME	REIGN
	Governor-General Adrienne Clarkson	1999–
Central African Republic	Prime Minister Elizabeth Domitien	1974–1976
Cherokee Nation	Chief Wilma Mankiller	1985–1995
China	Empress Wu Chao	655–705
	Dowager Empress Tsu-Hsi	1861–1908
	Dowager Empress Longyu	1911–1912
Denmark	Queen Margarethe I	1387–1412
	Queen Margrethe II 1972–	
Dominica	Prime Minister Mary Eugenia Charles	1980–1995
Easter Island	Paramount Chief Koreto Puakurunga	1868–1869?
	Paramount Chief Carolina	1869?–1888?
Egypt	Queen Hatshepsut	1503–1482 B.C.
	Queen Tiye	c. 1400 B.C.
	Queen Nefertiti	c. 1500 B.C.
	Queen Arsinoe II (joint ruler)	316–270 B.C.
	Queen Berenice III	80 B.C.
	Queen Cleopatra VII	51–30 B.C.
Ethiopia	Empress Candace	332 B.C.
	Empress Zauditu	1916–1930
Faeroe Islands	Prime Minister Marita Petersen	1993–1994
Finland	President Tarja Halonen	2000–
France	Prime Minister Edith Cresson	1991–1992
Georgia	Queen Tamara	1184–1212
Ghana	Queen Mother Yaa Asantewa	1863–1923
Great Britain	(UK) Boadicea, Queen of the Iceni	60–61
	Queen Jane (Lady Jane Grey)	1553 (9 days)
	Queen Mary I	1553–1558
	Queen Elizabeth I	1558–1603
	Queen Mary II (joint ruler)	1689–1702
	Queen Anne	1702–1714
	Queen Victoria	1837–1901
	Queen Elizabeth II	1952–
	Prime Minister Margaret Thatcher	1979–1990
Grenada	Governor Dame Hilda Louisa Bynoe	1967–1972
Guyana	Prime Minister Janet Jagan	1997
	President Janet Jagan	1997–1999

COUNTRY	NAME	REIGN
Haiti	Provisional President Ertha Pascal-Trouillot	1990, 1991
	Prime Minister Claudette Werleigh	1995–1996
Hawaii	Queen Liliuokalani	1891–1893
Hungary	Queen Mary	1382–1387
	Queen Elizabeth	1439–1440
	Queen Maria Theresa	1740–1780
Iceland	President Vigdis Finnbogadóttir	1980–1996
India	Prime Minister Indira Gandhi	1966–1977, 1980–1984
Indonesia	President Megawati Sukarnoputri	2001–
Ireland	President Mary Robinson	1990–1997
	President Mary McAleese	1997–
Israel	Prime Minister Golda Meir	1969–1974
Israel and Judah	Queen Athaliah	842–837 B.C.
Italy	Queen Theodelinda	590
	Queen Joanna I of Naples	1343–1381
	Queen Maria of Sicily	1377–1402
	Queen Joanna II of Naples	1414–1435
Japan	Empress Suiko Tenno	593–628
	Empress Kogyoku	642–645
	Empress Jito	686–697
	Empress Gemmyo	703–724
	Empress Koken (abdicated)	749–758
	Empress Shotuku-Koken	764–770
	Empress Toshi-ko	1762–1771
Latvia	President Vaira Vike-Freiberga	1999–
Lesotho	Paramount Chief 'Mantsebo Amelia 'Matsaba Sempe	1941–1960
Lithuania	Prime Minister Kazimiera Prunskiene	1990–1991
Luxembourg	Grand Duchess Marie Anne de Bragance (regent)	1908–1912
	Grand Duchess Marie-Adélaïde	1912–1919
	Grand Duchess Charlotte	1919–1964
Madagascar	Queen Ranavalona I	1828–1861
	Queen Rasoaherina	1863–1868
	Queen Ranavalona II	1868–1883
	Queen Ranavalona III (deposed)	1883–1897

COUNTRY	NAME	REIGN
Maldives	Amina Rani Kilagefanu	1757–1759
Malta	President Agatha Barbara	1982–1987
Micronesia	High Commissioner of Trust Territory of the Pacific Islands Janet J. McCoy	1981–1986
Monaco	Princess Louise-Hippolyte	1731
Netherlands	Queen Wilhelmina (abdicated)	1890–1948
	Queen Juliana	1948–1980
	Queen Beatrix	1980–
Netherlands Antilles	Prime Minister Lucinda da Costa Gomez-Matheeuws	1977
	Prime Minister Maria Liberia-Peters	1984–1986, 1988–1993
	Prime Minister Susanne Camelia-Romer	1993, 1998–1999
New Zealand	Governor-General Dame Catherine Tizard	1990–1996
	Governor-General Dame Silvia Cartwright	2001–
	Prime Minister Jenny Shipley	1997–1999
	Prime Minister Helen Clark	1999–
New Zealand (Maori community)	Queen Te Ata-I Rangi-Kahu Koroki Te Rata Mahuta	1966–
Nicaragua	President Maria Liberia Peres	1984–1985
	President Violeta Barriosde Chamorro	1990–1997
Nigeria	Queen Amina of Zaria	1588–1589
Norway	Queen Margaret	1387–1412
	Prime Minister Gro Harlem Brundtlandt	1981, 1986–1989, 1990–1996
Pakistan	Prime Minister Benazir Bhutto	1988–1990, 1993–1996
Panama	President Mireya Moscoso	1999–
Philippines	President Maria Corazon Aquino	1986–1992
	President Gloria Macapagal-Arroyo	2001–
Poland	Queen Hedwige	1384–1399
	Premier Hanna Suchocka	1992–1993
Portugal	Queen Maria I	1777–1816

COUNTRY	NAME	REIGN
	Queen Maria II	1826–1828, 1834–1853
	Prime Minister Maria de Lourdes Pintasilgo	1979 (149 days)
Roman Empire	Empress Irene	797–802
Russia	Empress Catherine I	1725–1727
	Empress Anna Ivanovna	1730–1740
	Empress Elizabeth Petrovna	1741–1762
	Empress Catherine II (Catherine the Great)	1762–1796
Rwanda	Prime Minister Agathe Uwilingiyimana	1993–1994
St. Lucia	Governor-General Dame Pearlette Louisy	1997–
Scotland	Queen Mary Stuart (executed)	1542–1567
Seminole Nation	Chief Betty Mae Jumper	1960–1969
Senegal	Mame Madior Boye	2001–
Sheba	Queen Makeda	960 B.C.
Spain	Queen Dona Urraca	1109–1126
	Queen Juana I	1274–1307
	Queen Juana II	1328–1349
	Queen Dona Blanca	1425–1441
	Queen Isabella I (joint ruler)	1474–1504
	Queen Isabella II	1833–1868
Sri Lanka	Queen Anula	47–42 B.C.
	Queen Sivali	35 B.C.
	Queen Lilavati	1197–1200, 1209–1212
	Queen Kalyanavati	1202–1208
	Prime Minister Sirimavo Bandaranaike	1960–1965, 1970–1977, 1994–2000
	President Chandrika Kumaratunga	1994–
Sudan	Queen Amanirenas of Kush	late first century B.C.
Sweden	Queen Christina (abdicated)	1632–1654
	Queen Ulrica Eleonora (abdicated)	1718–1720
Switzerland	President Ruth Dreifuss	1999
Tonga	Chief Tupoumahe'ofo	17??–1793

COUNTRY	NAME	REIGN
	Queen Salote Tubou III	1918–1965
Turkey	Prime Minister Tansu Çiller	1993–1996
Wallis	Ruler Toifale	1825
	Ruler Falakika Seilala Lavelua	1858–1869
	Ruler Amelia Tokagahahau Aliki Lavelua	1869–1895
	Ruler Aloisia Lavelua	1953–1958
Yugoslavia	Premier Milka Planinc	1982–1996
Zululand	Queen Nandi	1778–1826

MODERN QUEENS

In 2002, there were six reigning queens in Europe. Although they were the royalty in their countries, in all cases actual power was held by the national governments, which were run separately.

Queen Elizabeth II — United Kingdom
Queen Sofia — Spain
Queen Beatrix — The Netherlands
Queen Margrethe II — Denmark
Queen Silvia — Sweden
Queen Fabiola — Belgium

☞ SHE SAID IT ☜

"You can't shake hands with a clenched fist."

> — Indira Gandhi, former
> prime minister of India

"If you are guided by opinion polls, you are not practicing leadership — you are practicing follower-ship."

> — Margaret Thatcher, former prime
> minister of the United Kingdom

"I can honestly say that I was never affected by the question of the success of an undertaking. If I felt it was the right thing to do, I was for it regardless of the possible outcome."

> — Golda Meir, former prime minister of Israel

"I don't have any formula for ousting a dictator or building democracy. All I can suggest is to forget about yourself and just think of your people. It's always the people who make things happen."

> — Corazon Aquino, former
> president of the Philippines

"Who knows, somewhere out there in the audience may even be someone who will one day follow in my footsteps and preside over the White House as the president's spouse. I wish him well."

> — Barbara Bush, former first lady of the
> United States, in a commencement address